INTRODUCTORY STATISTICS AND ANALYTICS

INTRODUCTORY STATISTICS AND ANALYTICS

A Resampling Perspective

PETER C. BRUCE
Institute for Statistics Education
Statistics.com
Arlington, VA

Published by John Wiley & Sons, Inc., Hoboken, New Jersey
Published simultaneously in Canada

For general information on our other products and services or for technical support, please contact our Customer Care Department within the United States at (800) 762-2974, outside the United States at (317) 572-3993 or fax (317) 572-4002.

Wiley also publishes its books in a variety of electronic formats. Some content that appears in print may not be available in electronic formats. For more information about Wiley products, visit our web site at www.wiley.com.

Library of Congress Cataloging-in-Publication Data is available.

ISBN: 978-1-118-88135-4

Printed in the United States of America

CONTENTS

Preface   ix

Acknowledgments   xi

Introduction   xiii

1 Designing and Carrying Out a Statistical Study   1

    1.1   A Small Example, 3
    1.2   Is Chance Responsible? The Foundation of Hypothesis Testing, 3
    1.3   A Major Example, 7
    1.4   Designing an Experiment, 8
    1.5   What to Measure—Central Location, 13
    1.6   What to Measure—Variability, 16
    1.7   What to Measure—Distance (Nearness), 19
    1.8   Test Statistic, 21
    1.9   The Data, 22
    1.10 Variables and Their Flavors, 28
    1.11 Examining and Displaying the Data, 31
    1.12 Are we Sure we Made a Difference? 39
           Appendix: Historical Note, 39
    1.13 Exercises, 40

2 Statistical Inference   45

    2.1   Repeating the Experiment, 46
    2.2   How Many Reshuffles? 48
    2.3   How Odd is Odd? 53
    2.4   Statistical and Practical Significance, 55

2.5   When to use Hypothesis Tests, 56
2.6   Exercises, 56

3   Displaying and Exploring Data   **59**

3.1   Bar Charts, 59
3.2   Pie Charts, 61
3.3   Misuse of Graphs, 62
3.4   Indexing, 64
3.5   Exercises, 68

4   Probability   **71**

4.1   Mendel's Peas, 72
4.2   Simple Probability, 73
4.3   Random Variables and their Probability Distributions, 77
4.4   The Normal Distribution, 80
4.5   Exercises, 84

5   Relationship between Two Categorical Variables   **87**

5.1   Two-Way Tables, 87
5.2   Comparing Proportions, 90
5.3   More Probability, 92
5.4   From Conditional Probabilities to Bayesian Estimates, 95
5.5   Independence, 97
5.6   Exploratory Data Analysis (EDA), 99
5.7   Exercises, 100

6   Surveys and Sampling   **104**

6.1   Simple Random Samples, 105
6.2   Margin of Error: Sampling Distribution for a Proportion, 109
6.3   Sampling Distribution for a Mean, 111
6.4   A Shortcut—the Bootstrap, 113
6.5   Beyond Simple Random Sampling, 117
6.6   Absolute Versus Relative Sample Size, 120
6.7   Exercises, 120

7   Confidence Intervals   **124**

7.1   Point Estimates, 124
7.2   Interval Estimates (Confidence Intervals), 125
7.3   Confidence Interval for a Mean, 126
7.4   Formula-Based Counterparts to the Bootstrap, 126
7.5   Standard Error, 132
7.6   Confidence Intervals for a Single Proportion, 133
7.7   Confidence Interval for a Difference in Means, 136
7.8   Confidence Interval for a Difference in Proportions, 139

7.9   Recapping, 140
         Appendix A: More on the Bootstrap, 141
         Resampling Procedure—Parametric Bootstrap, 141
         Formulas and the Parametric Bootstrap, 144
         Appendix B: Alternative Populations, 144
         Appendix C: Binomial Formula Procedure, 144
7.10   Exercises, 147

8   Hypothesis Tests   **151**

8.1   Review of Terminology, 151
8.2   A–B Tests: The Two Sample Comparison, 154
8.3   Comparing Two Means, 156
8.4   Comparing Two Proportions, 157
8.5   Formula-Based Alternative—t-Test for Means, 159
8.6   The Null and Alternative Hypotheses, 160
8.7   Paired Comparisons, 163
         Appendix A: Confidence Intervals Versus Hypothesis Tests, 167
         Confidence Interval, 168
         Relationship Between the Hypothesis Test and the Confidence Interval, 169
         Comment, 170
         Appendix B: Formula-Based Variations of Two-Sample Tests, 170
         Z-Test With Known Population Variance, 170
         Pooled Versus Separate Variances, 171
         Formula-Based Alternative: Z-Test for Proportions, 172
8.8   Exercises, 172

9   Hypothesis Testing—2   **178**

9.1   A Single Proportion, 178
9.2   A Single Mean, 180
9.3   More Than Two Categories or Samples, 181
9.4   Continuous Data, 187
9.5   Goodness-of-Fit, 187
         Appendix: Normal Approximation; Hypothesis Test of a Single
         Proportion, 190
         Confidence Interval for a Mean, 190
9.6   Exercises, 191

10 Correlation   **193**

10.1   Example: Delta Wire, 194
10.2   Example: Cotton Dust and Lung Disease, 195
10.3   The Vector Product and Sum Test, 196
10.4   Correlation Coefficient, 199
10.5   Other Forms of Association, 204
10.6   Correlation is not Causation, 205
10.7   Exercises, 206

11 Regression   **209**

    11.1   Finding the Regression Line by Eye, 210
    11.2   Finding the Regression Line by Minimizing Residuals, 212
    11.3   Linear Relationships, 213
    11.4   Inference for Regression, 217
    11.5   Exercises, 221

12 Analysis of Variance—ANOVA   **224**

    12.1   Comparing More Than Two Groups: ANOVA, 225
    12.2   The Problem of Multiple Inference, 228
    12.3   A Single Test, 229
    12.4   Components of Variance, 230
    12.5   Two-Way ANOVA, 240
    12.6   Factorial Design, 246
    12.7   Exercises, 248

13 Multiple Regression   **251**

    13.1   Regression as Explanation, 252
    13.2   Simple Linear Regression—Explore the Data First, 253
    13.3   More Independent Variables, 257
    13.4   Model Assessment and Inference, 261
    13.5   Assumptions, 267
    13.6   Interaction, Again, 270
    13.7   Regression for Prediction, 272
    13.8   Exercises, 277

Index   **283**

PREFACE

This book was developed by Statistics.com to meet the needs of its introductory students, based on experience in teaching introductory statistics online since 2003. The field of statistics education has been in ferment for several decades. With this book, which continues to evolve, we attempt to capture three important strands of recent thinking:

1. Connection with the field of *data science*—an amalgam of traditional statistics, newer machine learning techniques, database methodology, and computer programming to serve the needs of large organizations seeking to extract value from "big data."

2. Guidelines for the introductory statistics course, developed in 2005 by a group of noted statistics educators with funding from the American Statistical Association. These Guidelines for Assessment and Instruction in Statistics Education (GAISE) call for the use of real data with active learning, stress statistical literacy and understanding over memorization of formulas, and require the use of software to develop concepts and analyze data.

3. The use of resampling/simulation methods to develop the underpinnings of statistical inference (the most difficult topic in an introductory course) in a transparent and understandable manner.

We start off with some examples of statistics in action (including two of statistics gone wrong) and then dive right in to look at the proper design of studies and account for the possible role of chance. All the standard topics of introductory statistics are here (probability, descriptive statistics, inference, sampling, correlation, etc.), but sometimes, they are introduced not as separate standalone topics but rather in the context of the situation in which they are needed.

Throughout the book, you will see "Try It Yourself" exercises. The answers to these exercises are found at the end of each chapter after the homework exercises.

BOOK WEBSITE

Data sets, Excel worksheets, software information, and instructor resources are available at the book website: www.introductorystatistics.com

PETER C. BRUCE

ACKNOWLEDGMENTS

STAN BLANK

The programmer for Resampling Stats, Stan has participated actively in many sessions of Statistics.com courses based on this work and has contributed well both to the presentation of regression and to the clarification and improvement of sections that deal with computational matters.

MICHELLE EVERSON

Michelle Everson, editor (2013) of the *Journal of Statistics Education*, has taught many sessions of the introductory sequence at Statistics.com and is responsible for the material on decomposition in the ANOVA chapter. Her active participation in the statistics education community has been an asset as we have strived to improve and perfect this text.

ROBERT HAYDEN

Robert Hayden has taught early sessions of this course and has written course materials that served as the seed from which this text grew. He was instrumental in getting this project launched.

In the beginning, Julian Simon, an early resampling pioneer, first kindled my interest in statistics with his permutation and bootstrap approach to statistics, his Resampling Stats software (first released in the late 1970s), and his statistics text on the same subject. Simon, described as an "iconoclastic polymath" by Peter Hall in his "Prehistory of the Bootstrap," (*Statistical Science*, 2003, vol. 18, #2), is the intellectual forefather of this work.

Our Advisory Board—Chris Malone, William Peterson, and Jeff Witmer (all active in GAISE and the statistics education community in general) reviewed the overall concept and outline of this text and offered valuable advice.

Thanks go also to George Cobb, who encouraged me to proceed with this project and reinforced my inclination to embed resampling and simulation more thoroughly than what is found in typical college textbooks.

Meena Badade also teaches using this text and has also been very helpful in bringing to my attention errors and points requiring clarification and has helped to add the sections dealing with standard statistical formulas.

Kuber Deokar, Instructional Operations Supervisor at Statistics.com, and Valerie Troiano, the Registrar at STatisticscom, diligently and carefully shepherded the use of earlier versions of this text in courses at Statistics.com.

The National Science Foundation provided support for the Urn Sampler project, which evolved into the Box Sampler software used both in this course and for its early web versions. Nitin Patel, at Cytel Software Corporation, provided invaluable support and design assistance for this work. Marvin Zelen, an early advocate of urn-sampling models for instruction, shared illustrations that sharpened and clarified my thinking.

Many students at The Institute for Statistics Education at Statistics.com have helped me clarify confusing points and refine this book over the years.

Finally, many thanks to Stephen Quigley and the team at Wiley, who encouraged me and moved quickly on this project to bring it to fruition.

INTRODUCTION

 As of the writing of this book, the fields of statistics and data science are evolving rapidly to meet the changing needs of business, government, and research organizations. It is an oversimplification, but still useful, to think of two distinct communities as you proceed through the book:

1. The traditional academic and medical *research communities* that typically conduct extended research projects adhering to rigorous regulatory or publication standards, and

2. Business and large organizations that use statistical methods to extract value from their data, often on the fly. Reliability and value are more important than academic rigor to this *data science community*.

IF YOU CAN'T MEASURE IT, YOU CAN'T MANAGE IT

You may be familiar with this phrase or its cousin: if you can't measure it, you can't fix it. The two come up frequently in the context of Total Quality Management or Continuous Improvement programs in organizations. The flip side of these expressions is the fact that if you do measure something and make the measurements available to decision-makers, the something that you measure is likely to change.

Toyota found that placing a real-time gas-mileage gauge on the dashboard got people thinking about their driving habits and how they relate to gas consumption. As a result, their gas mileage—miles they drove per gallon of gas—improved.

In 2003, the Food and Drug Administration began requiring that food manufacturers include trans fat quantities on their food labels. In 2008, it was found from a study that blood levels of trans fats in the population had dropped 58% since 2000 (reported in the *Washington Post*, February 9, 2012, A3).

Thus, the very act of measurement is, in itself, a change agent. Moreover, measurements of all sorts abound—so much so that the term Big Data came into vogue in 2011 to describe the huge quantities of data that organizations are now generating.

Big Data: If You Can Quantify and Harness It, You Can Use It

In 2010, a statistician from Target described how the company used customer transaction data to make educated guesses about whether customers were pregnant or not. On the strength of these guesses, Target sent out advertising flyers to likely prospects, centered around the needs of pregnant women.

How did Target use data to make those guesses? The key was data used to "train" a statistical model: data in which the outcome of interest—pregnant/not pregnant—was known in advance. Where did Target get such data? The "not pregnant" data was easy—the vast majority of customers were not pregnant so the data on their purchases was easy to come by. The "pregnant" data came from a baby shower registry. Both datasets were quite large, containing lists of items purchased by thousands of customers.

Some clues are obvious—purchase of a crib and baby clothes is a dead giveaway. But, from Target's perspective, by the time a customer purchases these obvious big ticket items, it was too late—they had already chosen their shopping venue. Target wanted to reach customers earlier, before they decided where to do their shopping for the big day. For that, Target used statistical modeling to make use of nonobvious patterns in the data that distinguish pregnant from nonpregnant customers. One such clue was shifts in the pattern of supplement purchases—for example, a customer who was not buying supplements 60 days ago but is buying them now. Crafting a marketing campaign on the basis of educated guesses about whether a customer is pregnant aroused controversy for Target, needless to say.

Much of the book that follows deals with important issues that can determine whether data yields meaningful information or not:

- The role that random chance plays in creating apparently interesting results or patterns in data.
- How to design experiments and surveys to get useful and reliable information.
- How to formulate simple statistical models to describe relationships between one variable and another.

PHANTOM PROTECTION FROM VITAMIN E

In 1993, researchers examining a database on nurses' health found that nurses who took vitamin E supplements had 30–40% fewer heart attacks than those who did not. These data fit with theories that antioxidants such as vitamins E and C could slow damaging processes within the body. Linus Pauling, winner of the Nobel Prize in Chemistry in 1954, was a major proponent of these theories. The Linus Pauling Institute at Oregon State University is still actively promoting the role of vitamin E and other nutritional supplements in inhibiting disease. These results provided a major boost to the dietary supplements industry. The only problem? The heart health benefits of vitamin E turned out to be illusory. A study completed in 2007 divided 14,641 male physicians randomly into four groups:

1. Take 400 IU of vitamin E every other day
2. Take 500 mg of vitamin C every day
3. Take both vitamin E and C
4. Take placebo.

Those who took vitamin E fared no better than those who did not take vitamin E. As the only difference between the two groups was whether or not they took vitamin E, if there were a vitamin E effect, it would have shown up. Several meta-analyses, which are consolidated reviews of the results of multiple published studies, have reached the same conclusion. One found that vitamin E at the above dosage might even increase mortality.

What made the researchers in 1993 think that they had found a link between vitamin E and disease inhibition? After reviewing a vast quantity of data, researchers thought that they saw an interesting association. In retrospect, with the benefit of a well-designed experiment, it appears that this association was merely a chance coincidence. Unfortunately, coincidences happen all the time in life. In fact, they happen to a greater extent than we think possible.

STATISTICIAN, HEAL THYSELF

In 1993, Mathsoft Corp., the developer of Mathcad mathematical software, acquired StatSci, the developer of S-PLUS statistical software, predecessor to the open-source R software. Mathcad was an affordable tool popular with engineers—prices were in the hundreds of dollars, and the number of users was in the hundreds of thousands. S-PLUS was a high-end graphical and statistical tool used primarily by statisticians—prices were in the thousands of dollars, and the number of users was in the thousands.

In an attempt to boost revenues, Mathsoft turned to an established marketing principle—cross-selling. In other words, trying to convince the people who bought product A to buy product B. With the acquisition of a highly regarded niche product, S-PLUS, and an existing large customer base for Mathcad, Mathsoft decided that the logical thing to do would be to ramp up S-PLUS sales via direct mail to its installed Mathcad user base. It also decided to purchase lists of similar prospective customers for both Mathcad and S-PLUS.

This major mailing program boosted revenues, but it boosted expenses even more. The company lost over $13 million in 1993 and 1994 combined—significant numbers for a company that had only $11 million in revenue in 1992.

What Happened?

In retrospect, it was clear that the mailings were not well targeted. The costs of the unopened mail exceeded the revenue from the few recipients who did respond. In particular, Mathcad users turned out to be unlikely users of S-PLUS. The huge losses could have been avoided through the use of two common statistical techniques:

1. Doing a test mailing to the various lists being considered to (a) determine whether the list is productive and (b) test different headlines, copy, pricing, and so on, to see what works best.

2. Using predictive modeling techniques to identify which names on a list are most likely to turn into customers.

IDENTIFYING TERRORISTS IN AIRPORTS

Since the September 11, 2001 Al Qaeda attacks in the United States and subsequent attacks elsewhere, security screening programs at airports have become a major undertaking, costing billions of dollars per year in the United States alone. Most of these resources are consumed by an exhaustive screening process. All passengers and their tickets are reviewed, their baggage is screened, and individuals pass through detectors of varying sophistication. An individual and his or her bag can only receive a limited amount of attention in an exhaustive screening process. The process is largely the same for each individual. Potential terrorists can see the process and its workings in detail and identify its weaknesses.

To improve the effectiveness of the system, security officials have studied ways of focusing more concentrated attention on a small number of travelers. In the years after the attacks, one technique enhanced the screening for a limited number of randomly selected travelers. Although it adds some uncertainty to the process, which acts as a deterrent to attackers, random selection does nothing to focus attention on high-risk individuals.

Determining who is of high risk is, of course, the problem. How do you know who the high-risk passengers are?

One method is passenger profiling—specifying some guidelines about what passenger characteristics merit special attention. These characteristics were determined by a reasoned, logical approach. For example, purchasing a ticket for cash, as the 2001 hijackers did, raises a red flag. The Transportation Security Administration trains a cadre of Behavior Detection Officers. The Administration also maintains a specific no-fly list of individuals who trigger special screening.

There are several problems with the profiling and no-fly approaches.

- Profiling can generate backlash and controversy because it comes close to stereotyping. American National Public Radio commentator Juan Williams was fired when he made an offhand comment to the effect that he would be nervous about boarding an aircraft in the company of people in full Muslim garb.
- Profiling, as it does tend to merge with stereotype and is based on logic and reason, enables terrorist organizations to engineer attackers who do not meet profile criteria.
- No-fly lists are imprecise (a name may match thousands of individuals) and often erroneous. Senator Edward Kennedy was once pulled aside because he supposedly showed up on a no-fly list.

An alternative or supplemental approach is a statistical one—separate out passengers who are "different" for additional screening, where "different" is defined quantitatively across many variables that are not made known to the public. The statistical term is "outlier." Different does not necessarily prove that the person is a terrorist threat, but the theory is that outliers may have a higher threat probability. Turning the work over to a statistical algorithm mitigates some of the controversy around profiling as security officers would lack the authority to make discretionary decisions.

Defining "different" requires a statistical measure of distance, which we will learn more about later.

LOOKING AHEAD IN THE BOOK

We will be studying many things, but several important themes will be the following:

1. Learning more about random processes and statistical tools that will help quantify the role of chance and distinguish real phenomena from chance coincidence.
2. Learning how to design experiments and studies that can provide more definitive answers to questions such as whether vitamin E affects heart attack rates and whether to undertake a major direct mail campaign.
3. Learning how to specify and interpret statistical models that describe the relationship between two variables or between a response variable and several "predictor" variables, in order to
   - explain/understand phenomena and answer research questions ("Does a new drug work?" "Which offer generates more revenue?")
   - make predictions ("Will a given subscriber leave this year?" "Is a given insurance claim fraudulent?")

RESAMPLING

An important tool will be resampling—the process of taking repeated samples from observed data (or shuffling that data) to assess what effect random variation might have on our statistical estimates, our models, and our conclusions. Resampling was present in the early days of statistical science, but, in the absence of computers, was quickly superceded by formula approaches. It has enjoyed a resurgence in the last 30 years.

Resampling in Data Mining: Target Shuffling

John Elder is the founder of the data mining and predictive analytics services firm Elder Research. He tests the accuracy of his data mining results through a process he calls "target shuffling". It's a method Elder says is particularly useful for identifying false positives, or when events are perceived to have a cause-and-effect relationship, as opposed to a coincidental one.

"The more variables you have, the easier it becomes to "over-search" and identify (false) patterns among them," Elder says—what he calls the 'vast search effect.'

As an example, he points to the Redskins Rule, where for over 70 years, if the Washington Redskins won their last home football game, the incumbent party would win the presidential election. "There's no real relationship between those two things," Elder says, "but for generations, they just happened to line up."

As hypotheses generated by automated search grow in number, it becomes easy to make inferences that are not only incorrect, but dangerously misleading. To prevent this problem, Elder Research uses target shuffling with all of their clients. It reveals how likely it is that results as strong as you found could have occurred by chance.

"Target shuffling is a computer simulation that does what statistical tests were designed to when they were first invented," Elder explains. "But this method is much easier to understand, explain, and use than those mathematical formulas."

Here's how the process works. On a set of training data:

1. Build a model to predict the target variable (output) and note its strength (e.g., R-squared, lift, correlation, explanatory power).
2. Randomly shuffle the target vector to "break the relationship" between each output and its vector of inputs.
3. Search for a new best model – or "most interesting result" - and save its strength. (It is not necessary to save the model; its details are meaningless by design.)
4. Repeat steps 2 and 3 many times and create a distribution of the strengths of all the bogus "most interesting" models or findings.
5. Evaluate where your actual results (from step 1) stand on (or beyond) this distribution. This is your "significance" measure or probability that a result as strong as your initial model can occur by chance.

Let's break this down: imagine you have a math class full of students who are going to take a quiz. Before the quiz, everyone fills out a card with specified personal information, such as name, age, how many siblings they have, and what other math classes they've taken. Everyone then takes the quiz and receives their score.

To discover why certain students scored higher than others, you could model the target variable (the grade each student received) as a function of the inputs (students' personal information) to identify patterns. Let's say you find that older sisters have the highest quiz scores, which you think is a solid predictor of which types of future students will perform the best.

But depending on the size of the class and the number of questions you asked everyone, there's always a chance that this relationship is not real, and therefore won't hold true for the next class of students. (Even if the model seems reasonable, and facts and theory can be brought to support it, the danger of being fooled remains: "Every model finding seems to cause our brains to latch onto corroborating explanations instead of generating the critical alternative hypotheses we really need.")

With target shuffling, you compare the same inputs and outputs against each other a second time to test the validity of the relationship. This time, however, you randomly shuffle the outputs so each student receives a different quiz score—Suzy gets Bob's, Bob gets Emily's, and so forth.

All of the inputs (personal information) remain the same for each person, but each now has a different output (test score) assigned to them. This effectively breaks the relationship between the inputs and the outputs without otherwise changing the data.

You then repeat this shuffling process over and over (perhaps 1000 times, though even 5 times can be very helpful), comparing the inputs with the randomly assigned outputs each time. While there should be no real relationship between each student's personal information and these new, randomly assigned test scores, you'll inevitably find some new false positives or "bogus" relationships (e.g. older males receive the highest scores, women who also took Calculus receive the highest scores, etc.).

As you repeat the process, you record these "bogus" results over the course of the 1000 random shufflings. You then have a comparison distribution that you can use to assess whether the result that you observed in reality is truly interesting and impressive or to what degree it falls in the category of "might have happened by chance."

Elder first came up with target shuffling 20 years ago, when his firm was working with a client who wasn't sure if he wanted to invest more money into a new hedge fund. While the fund had done very well in its first year, it had been a volatile ride, and the client was unsure if the success was due to luck or skill. A standard statistical test showed that the probability of the fund being that successful in a chance model was very low, but the client wasn't convinced.

So Elder performed 1,000 simulations where he shuffled the results (as described above) where the target variable was the daily buy or hold signal for the next day. He then compared the random results to how the hedge fund had actually performed.

Out of 1,000 simulations, the random distribution returned better results in just 15 instances—in other words, there was only a 1.5% chance that the hedge fund's success could occur just as the result of luck. This new way of presenting the data made sense to the client, and as a result he invested 10 times as much in the fund.[1]

"I learned two lessons from that experience," Elder says. "One is that target shuffling is a very good way to test non-traditional statistical problems. But more importantly, it's a process that makes sense to a decision maker. Statistics is not persuasive to most people—it's just too complex.

"If you're a business person, you want to make decisions based upon things that are real and will hold up. So when you simulate a scenario like this, it quantifies how likely it is that the results you observed could have arisen by chance in a way that people can understand."

BIG DATA AND STATISTICIANS

Before the turn of the millennium, by and large, statisticians did not have to be too concerned with programming languages, SQL queries, and the management of data. Database administration and data storage in general was someone else's job, and statisticians would obtain or get handed data to work on and analyze. A statistician might, for example,

- Direct the design of a clinical trial to determine the efficacy of a new therapy
- Help a psychology student determine how many subjects to enroll in a study
- Analyze data to prepare for legal testimony
- Conduct sample surveys and analyze the results
- Help a scientist analyze data that comes out of a study
- Help an engineer improve an industrial process

[1] The fund went on to do well for a decade; the story is recounted in chapter 1 of Eric Siegel's *Predictive Analytics* (Wiley, 2013)

All of these tasks involve examining data, but the number of records is likely to be in the hundreds or thousands at most, and the challenge of obtaining the data and preparing it for analysis was not overwhelming. So the task of obtaining the data could safely be left to others.

Data Scientists

The advent of big data has changed things. The explosion of data means that more interesting things can be done with data, and they are often done in real time or on a rapid turnaround schedule. FICO, the credit-scoring company, uses statistical models to predict credit card fraud, collecting customer data, merchant data, and transaction data 24 hours a day. FICO has more than two billion customer accounts to protect, so it is easy to see that this statistical modeling is a massive undertaking.

Computer programming and database administration lie beyond the scope of this course but not beyond the scope of statistical studies. See the book website for links to over 100 online courses, to get an idea of what statistics covers now. The statistician must be conversant with the data, and the data keeper now wants to learn the analytics:

- Statisticians are increasingly asked to plug their statistical models into big data environments, where the challenge of wrangling and preparing analyzable data is paramount, and requires both programming and database skills.
- Programmers and database administrators are increasingly interested in adding statistical methods to their toolkits, as companies realize that they have strategic, not just clerical value hidden in their databases.

Around 2010, the term *data scientist* came into use to describe analysts who combined these two sets of skills. Job announcements now carry the term *data scientist* with greater frequency than the term *statistician*, reflecting the importance that organizations attach to managing, manipulating, and obtaining value out of their vast and rapidly growing quantities of data.

We close with a probability experiment:

Try It Yourself 1.1

Let us look first at the idea of randomness via a classroom exercise.

1. Write down a series of 50 random coin flips without actually flipping the coins. That is, write down a series of 50 Hs and Ts selected in such a way that they appear random.
2. Now, actually flip a coin 50 times.

If you are reading this book in a course, please report your results to the class for compilation—specifically, report two lists of Hs and Ts like this: My results—Made up flips: HTHHHTT, and so on. Actual flips: TTHHTHTHTH, and so on.

1

DESIGNING AND CARRYING OUT A STATISTICAL STUDY

In this chapter we study random behavior and how it can fool us, and we learn how to design studies to gain useful and reliable information. After completing this chapter, you should be able to

- use coin flips to replicate random processes and interpret the results of coin-flipping experiments,
- define and understand probability,
- define, intuitively, p-value,
- list the key statistics used in the initial exploration and analysis of data,
- describe the different data formats that you will encounter, including relational database and flat file formats,
- describe the difference between data encountered in traditional statistical research and "big data,"
- explain the use of treatment and control groups in experiments,
- explain the role of randomization in assigning subjects in a study,
- explain the difference between observational studies and experiments.

You may already be familiar with statistics as a method of gathering and reporting data. Sports statistics are a good example of this. For many decades, data have been collected and reported on the performance of both teams and players using standard metrics such as yards via pass completions (quarterbacks in American football), points scored (basketball), and batting average (baseball).

Introductory Statistics and Analytics: A Resampling Perspective, First Edition. Peter C. Bruce.
© 2015 John Wiley & Sons, Inc. Published 2015 by John Wiley & Sons, Inc.

Sports fans, coaches, analysts, and administrators have a rich array of useful statistics at their disposal, more so than most businesses. TV broadcasters can not only tell you when a professional quarterback's last fumble was but they can also queue up television footage almost instantly, even if that footage dates from the player's college days. To appreciate the role that statistical analysis (also called *data analytics*) plays in the world today, one needs to look no further than the television broadcast of a favorite sport—pay close attention to the statistics that are reported and imagine how they are arrived at.

The whole point in sports, of course, is statistical—to score more points than the other player or the other team. The activities of most businesses and organizations are much more complex, and valid statistical conclusions are more difficult to draw, no matter how much data are available.

Big Data

In most organizations today, raw data are plentiful (often too plentiful), and this is a two-edged sword.

- Huge amounts of data make prediction possible in circumstances where small amounts of data do not help. One type of recommendation system, for example, needs to process large numbers of transactions to locate transactions with the same items you are looking at—enough so that reliable information about associated items can be deduced.
- On the other hand, huge data flows can obscure the signal, and useful data are often difficult and expensive to gather. We need to find ways to get the most information and the most accurate information for each dollar spent in gathering and preparing data.

Data Mining and Data Science

The terms *big data*, *data mining*, *data science*, and *predictive analytics* often go together, and when people think of data mining various things come to mind. Laypersons may think of large corporations or spy agencies combing through petabytes of personal data in hopes of locating tidbits of information that are interesting or useful. Analysts often consider data mining to be much the same as predictive analytics—training statistical models to use known values ("predictor variables") to predict an unknown value of interest (loan default, acceptance of a sales offer, filing a fraudulent insurance claim, or tax return).

In this book, we will focus more on standard research statistics, where data are small and well structured, leaving the mining of larger, more complex data to other books. However, we will offer frequent windows into the world of data science and data mining and point out the connections with the more traditional methods of statistics.

In any case, it is still true that most data science, when it is well practiced, is not just aimless trolling for patterns but starts out with questions of interest such as:

- What additional product should we recommend to a customer?
- Which price will generate more revenue?
- Does the MRI show a malignancy?
- Is a customer likely to terminate a subscription?

All these questions require some understanding of random behavior and all benefit from an understanding of the principles of well-designed statistical studies, so this is where we will start.

1.1 A SMALL EXAMPLE

In the fall of 2009, the Canadian Broadcasting Corporation (CBC) aired a radio news report on a study at a hospital in Quebec. The goal of the study was to reduce medical errors. The hospital instituted a new program in which staff members were encouraged to report any errors they made or saw being made. To accomplish that, the hospital agreed not to punish those who made errors. The news report was very enthusiastic and claimed that medical errors were less than half as common after the new program was begun. An almost parenthetical note at the end of the report mentioned that total errors had not changed much, but major errors had dropped from seven, the year before the plan was begun, to three, the year after (Table 1.1).

TABLE 1.1   Major Errors in a Quebec Hospital

Before no-fault reporting	Seven major errors
After no-fault reporting	Three major errors

1.2 IS CHANCE RESPONSIBLE? THE FOUNDATION OF HYPOTHESIS TESTING

This seems impressive, but a statistician recalling the vitamin E case might wonder if the change is real or if it could just be a fluke of chance. This is a common question in statistics and has been formalized by the practices and policies of two groups:

- Editors of thousands of journals who report the results of scientific research because they want to be sure that the results they publish are real and not chance occurrences.
- Regulatory authorities, mainly in medicine, who want to be sure that the effects of drugs, treatments, and so on are real and are not due to chance.

A standard approach exists for answering the question "is chance responsible?" This approach is called a *hypothesis test*. To conduct one, we first build a plausible mathematical model of what we mean by chance in the situation at hand. Then, we use that model to estimate how likely it is, just by chance, to get a result as impressive as our actual result. If we find that an impressive improvement like the observed outcome would be very unlikely to happen by chance, we are inclined to reject chance as the explanation. If our observed result seems quite possible according to our chance model, we conclude that chance is a reasonable explanation. We now conduct a hypothesis test for the Quebec hospital data.

What do we mean by the outcome being "just" chance? How should that chance model look like? We mean that there is nothing remarkable going on — that is, the no-fault reporting has no effect, and the $7 + 3 = 10$ major errors just happened to land seven in the first year and three in the second. If there is no treatment effect from no-fault reporting and only chance were operating, we might expect 50/50 or five in each year, but we would not *always* get five each year if the outcome were due to chance. One way that we could see what might happen would be to just toss a coin 10 times, letting the 10 tosses represent the 10 major errors, and letting heads represent the first year and tails the second. Then a toss of HTTHTTHHHH would represent six in the first year and four in the second.

Try It Yourself 1.1

Toss a coin 10 times and record the number of heads and the number of tails. We will call the 10 tosses one trial. Then repeat that trial 11 more times for a total of 12 trials and 120 tosses. To try this exercise on your computer, use the macro-enabled Excel workbook `boxsampler1.xlsm` (located at the book website), which contains a Box Sampler model.

The textbook supplements contain both Resampling Stats for Excel and StatCrunch procedures for this problem.

Did you ever get seven (or more) heads in a trial of 10 tosses? (Answers to "Try it Yourself" exercises are at the end of the chapter.)

Let us recap the building blocks of our model:

- A single coin flip, representing the allocation of a single error to this year (T in the above discussion) or the prior year (H in the above discussion).
- A series of 10 coin flips, representing a single simulation, also called a *trial*, that has the same sample size as the original sample of 10 errors.
- Twelve repetitions of that simulation.

At this stage, you have an initial impression of whether seven or more heads is a rare event. But you only did 12 trials. We picked 12 as an arbitrary number, just to get started. What is next?

One option is to sit down and figure out exactly what the probability is of getting seven heads, eight heads, nine heads, or 10 heads. Recall that our goal is to learn whether seven heads and only three tails are an extreme, that is, it is an unusual occurrence. If we get lots of cases where we get eight heads, nine heads, and so on, then clearly, seven heads is not extreme or unusual.

 Why do we count ≥ 7 instead of $=7$? This is an important but often a misunderstood point. If it is not clear, please raise it in class!

We have used the terms "*probability*" and "*chance*," and you probably have a good sense of what they mean, for example, probability of precipitation or chance of precipitation. Still, let us define them—the meaning is the same for each, but probability is a more specific statistical term so we will stick with that.

Definition: **A somewhat subjective definition of probability**

The probability of something happening is the proportion of time that it is expected to happen when the same process is repeated over and over (paraphrasing from Freedman, et al., *Statistics*, 2nd ed., Norton, 1991, 1st ed. 1978).

Definition: **Probability defined more like a recipe or formula**

First, turn the problem into a box filled with slips of paper, with each slip representing a possible outcome for an event. For example, a box of airline flights would have a label for each flight: late, on time, or canceled. The probability of an outcome is the number of slips of paper with that outcome divided by the total number of slips of paper in the box.

Question 1.1

From the above, particularly the second definition, you can see that the probability of something happening must always lie between _____ and _____, inclusive.

You can speak in terms of either proportions or percentages—40% is the same as 0.40.

Earlier, we calculated all the possible outcomes for three flips of a coin. Can we do the same thing for 10 flips? If you try it, you will see that this method of counting will quickly become tedious.

Three flips is easier—here is a video from the Khan Academy that illustrates how to calculate the probability of two heads in three tosses by counting up the possibilities.

https://www.youtube.com/watch?v=3UlE8gyKbkU&feature=player_embedded

With 10 flips, one option is to do many more simulations. We will get to that in a bit, but, for now, we will jump to the conclusion so that we can continue with the overall story.

The probability of getting seven or more heads is about $2/12 = 0.1667$.

Interpreting This Result

The value 0.1667 means that such an outcome, i.e., seven or more heads, is not all that unusual, and the results reported from Canada could well be due to chance. This news story was not repeated later that day nor did it appear on the CBC website, so perhaps they heard from a statistician and pulled the story.

Question 1.2

Would you consider chance as a reasonable explanation if there were 10 major errors the year before the change and none the year after? Hint: use the coin tosses that you already performed.

Suppose it had turned out the other way. If our chance model had given a very low probability to the actual outcome, then we are inclined to reject chance as the main factor.

Definition: **p-value**

If we examine the results of the chance model simulations in this way, the probability of seeing a result as extreme as the observed value is called the *p-value* (or probability value).

Even if our chance model had produced a very low probability, ruling out chance, this does not necessarily mean that the real cause is the new no-fault reporting policy. There are many other possible explanations. Just as we need to rule out chance, we need to rule out those as well. For example, we might be more impressed if our hospital was unique—reducing its errors while every other hospital in Quebec had more major errors the second year. Conversely, we would be less impressed if the number of errors went down at all hospitals that second year—including those with no new program.

Do not worry if this definition of *p*-value and the whole hypothesis testing process are not fully clear to you at this early stage. We will come back to it repeatedly.

The use of *p*-values is widespread; their use as decision-making criteria lies more in the *research* community than in the *data science* community.

Increasing the Sample Size

Intuition tells us that small samples lead to fluke results. Let us see what happens when we increase the sample size.

Try It Yourself 1.2

Let us double the sample size and imagine that the study had revealed 14 errors in 1 year and six the following, instead of seven and three. Now, regroup your 12 simulations of 10 tosses into six trials of 20 tosses each. Combine the first and the second sets, the third and fourth, and so on. Then do six more trials of 20 tosses each for a total of 120 additional tosses. You should now have 12 sets of 20 tosses. If you want to try a computer simulation, use the Box Sampler macro-enabled Excel workbook boxsampler2.xlsm.

The textbook supplements contain a Resampling Stats procedure for this problem. Did you ever get 14 or more heads in a trial of 20 tosses?

Technique

We will use the "Technique" heading for the details you need to do the analyses. We illustrate the use of a computer to generate random numbers, which is shown as follows.

 In our original example, we saw seven errors in the first year and three errors in the next, for a *reduction* of four errors. As we develop this example further, we will deal exclusively with data on *reduction* in errors.

Tossing coins can get tiresome and can only model events that have a 50/50 chance of either happening or not happening. Modeling random events is typically done by generating random numbers by computer.

Excel, for example, has two options for generating random numbers:

RAND generates a random number between 0 and 1.

RANDBETWEEN generates a random integer between two values that you specify.

You then need to map the random digit that is generated to the outcome of an event that you are trying to model. For example:

1. A customer of a music streaming subscription service has a 15% probability of canceling the service in a given year. This could be modeled by generating a random integer between 1 and 100 and labeling 1–15 as "cancel" and 16–100 as "maintain subscription."

In Excel, the function would be entered as =RANDBETWEEN(1,100).

After generating, say, 1000 random numbers (and putting them in cells A1:A1000), you could count the number of cancelations using COUNTIF:

=COUNTIF(A1:A1000,"<=15").

What is a Random Number?

For our purposes, we can think of a random number as the result of placing the digits 0–9 in a hat or a box, shuffling the hat or box, and then drawing a digit. Most random numbers are produced by computer algorithms that produce series of numbers that are effectively random and unpredictable, or at least sufficiently random for the purpose at hand. But the numbers are produced by an algorithm that is technically called a *pseudo-random number generator*. There have been many research studies and scholarly publications on the properties of random number generators (RNGs) and the computer algorithms they use to produce pseudo random numbers. Some are better than others; the details of how they work are beyond the scope of this book. We can simply think of random number generators as the computer equivalent of picking cards from a hat or a box that has been well shuffled.

1.3   A MAJOR EXAMPLE

To tie together our study of statistics, we will look at one major example. Using the study reported by the CBC as our starting point, we introduce basic but important statistical concepts.

Imagine that you have just been asked to design a better study to determine if the sort of no-fault accident reporting tried in a Quebec hospital really does reduce the number of serious medical errors. The standard type of study in such a situation would be an *experiment*.

Experiment versus Observational Study

In the fifth inning of the third game of the 1932 baseball World Series between the NY Yankees and the Chicago Cubs, the great slugger Babe Ruth came to bat and pointed toward center field as if to indicate that he planned to hit the next pitch there. On the next pitch, he indeed hit the ball for a home run into the centerfield bleachers.*

A Babe Ruth home run was an impressive feat but not that uncommon. He hit one every 11.8 at bats. What made this one so special is that he predicted it. In statistical terms, he specified in advance a theory about a future event—the next swing of the bat—and an outcome of interest—home run to centerfield.

In statistics, we make an important distinction between studying preexisting data—an observational study—and collecting data to answer a prespecified question—an experiment or a prospective study.

We will learn more about this later but keep in mind that the most impressive and durable results in science come when the researcher specifies a question in advance and then collects data in a well-designed experiment to answer the question. Offering commentary on the past can be helpful but is no match for predicting the future.

*There is some controversy about whether he actually pointed to center field or to left field and whether he was foreshadowing a prospective home run or taunting Cubs players. You can Google the incident ("Babe Ruth called shot") and study videos on YouTube and then judge for yourself.

1.4   DESIGNING AN EXPERIMENT

The principles of designing an experiment are fundamental and should be studied by both data scientists and research statisticians. When it comes to practice and implementation, this material will be of primary interest to the *research* community identified at the beginning of the introduction.

In our errors experiment, we could compare two groups of hospitals. One group uses the no-fault plan and the other does not. The group that gets the change in treatment you wish to study is called the *treatment group*. The group that gets no treatment or the standard treatment is called the *control group*. Normally, you need some reference group for comparison, although in some studies you may be comparing multiple treatments with no control. How do you decide which hospitals go into which group?

You would like the two groups to be similar to one another, except for the treatment/control difference. That way, if the treatment group does turn out to have fewer errors, you can be confident that it was due to the treatment. One way to do this would be to study all the hospitals in detail, examine all their relevant characteristics, and assign them to treatment/control in such a way that the two groups end up being similar across all these attributes. There are two problems with this approach.

1. It is usually not possible to think of all the relevant characteristics that might affect the outcome. Research is replete with the discovery of factors that were unknown prior to the study or thought to be unimportant.
2. The researcher, who has a stake in the outcome of the experiment, may consciously or unconsciously assign hospitals in a way that enhances the chances of the success of his or her pet theory.

Oddly enough, the best strategy is to assign hospitals randomly—perhaps by tossing a coin.

Randomizing

True random assignment eliminates both conscious and unconscious bias in the assignment to groups. It does not guarantee that the groups will be equal in all respects. However, it does guarantee that any departure from equality will be due simply to the chance allocation and that the larger the number of samples, the fewer differences the groups will have. With extremely large samples, differences due to chance virtually disappear and you are left with differences that are real—provided the assignment to groups is really random.

Law of Large Numbers

The law of large numbers states that, despite short-term average deviations from an event's theoretical mean, such as the chance of a coin landing heads, the long-run empirical—actual—average occurrence of the event will approach, with greater and greater precision, the theoretical mean. The short-run deviations get washed out in a flood of trials. During World War II, John Kerrich, a South African mathematician, was imprisoned in Denmark. In his idle moments, he conducted several probability experiments.

In one such experiment, he flipped a coin repeatedly, keeping track of the number of flips and the number of heads. After 20 flips, he was exactly even—10 heads and 10 tails. After 100 flips, he was down six heads—44 heads and 56 tails—or 6%. After 500 flips, he was up five heads—255 heads and 245 tails—or 1%. After 10,000 flips, he was up 67 heads or 0.67%.

A plot of all his results with the proportion of heads on the y-axis and the number of tosses on the x-axis shows a line that bounces around a lot on the left side but settles down to a straighter and straighter line on the right side, tending toward 50%.

 Do not confuse the Law of Large Numbers with the popular conception of the Law of Averages.

Law of Large Numbers

Long run actual average will approach the theoretical average.

Law of Averages

A vague term, sometimes meaning as mentioned earlier but also used popularly to refer to the mistaken belief that, after a string of heads, the coin is "due" to land tails, thereby preserving its 50/50 probability in the long run. One often encounters this concept in sports, for example, a batter is "due" for a hit after a dry spell.

Random assignment let us make the claim that any difference in the group outcomes that can *more* than might happen by chance is, in fact, due to the different treatment received by the groups. Kerrich had a lot of time on his hands and could accumulate a huge sample under controlled conditions for his simple problem. In actual studies, researchers rarely have the ability to collect samples sufficiently large that we can dismiss chance as a factor. In the study of probability in this course, lets us quantify the role that chance can play and take it into account (Figure 1.1).

Even if we performed a dummy experiment in which both groups got the same treatment, we would expect to see some differences from one hospital to another. An everyday example of this might be tossing a coin. You get different results from one toss to the next just by chance. Check the coin tosses you did earlier in connection with the CBC news report on

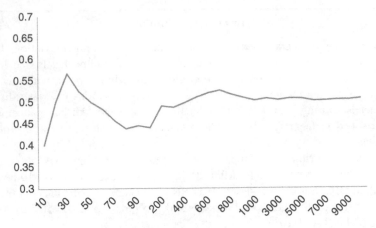

Figure 1.1   Kerrich coin tosses. Number of tosses on the *x*-axis and the proportion of heads on the *y*-axis.

medical errors. If you toss a coin 10 times, you will get a certain number of heads. Do it again and you will probably get a different number of heads.

Although the results vary, there are laws of chance that allow you to calculate things like how many heads you would expect on average or how much the results would vary from one set of 10 tosses to the next. If we assign subjects at random, we can use these same laws of chance—or a lot of coin tosses—to analyze our results.

If we have Doctor Jones assign subjects using her own best judgment, we will have no mathematical theory to guide us. That is because it is very unlikely that we can find any books on how Doctor Jones assigns hospitals to the treatment and control groups. However, we can find many books on random assignment. It is a standard, objective way of doing things that works the same for everybody. Unfortunately, it is not always possible. Human subjects can neither be assigned a gender nor a disease.

Planning

You need some hospitals and you estimate that you can find about 100 within reasonable distance. You will probably need to present a plan for your study to the hospitals to get their approval. This seems like a nuisance, but they cannot let just anyone do any study they please on the patients. Studies of new prescription drugs require government approval as well, which is a long and costly process. In addition to writing a plan to get approval, you know that one of the biggest problems in interpreting studies is that many are poorly designed. You want to avoid that so you think carefully about your plan and ask others for advice. It would be good to talk to a statistician who has experience in medical work. Your plan is to ask the 100 or so available hospitals if they are willing to join your study. They have the right to say no. You hope that quite a few will say yes. In particular, you hope to recruit 50 willing hospitals and randomly assign them to two groups of 25.

Try It Yourself 1.3

Suppose you wanted to study the impact of watching television on violent behavior with an experiment. What issues might you encounter in trying to assign treatments to subjects? What would the treatment be?

Blinding

We saw that randomization is used to try to make the two groups similar at the beginning. It is important to keep them as similar as possible. We want to be sure that the treatment is the only difference between them. One subtle difference we have to worry about when working with humans is that their behavior can be changed by the fact that they are participating in a study.

Out-of-Control Toyotas?

In the fall of 2009, the National Highway Transportation Safety Agency received several dozen complaints per month about Toyota cars speeding out of control. The rate of complaint was not that different from the rates of complaint for other car companies. Then, in November of 2009, Toyota recalled 3.8 million vehicles to check for sticking gas pedals. By February, the complaint rate had risen from several dozen per month to over 1500 per month of alleged cases of unintended acceleration. Attention turned to the electronic throttle.

Clearly, what changed was not the actual condition of cars—the stock of Toyotas on the road in February of 2010 was not that different from November of 2009. What changed was car owners' awareness and perception as a result of the headlines surrounding the recall. Acceleration problems, whether real or illusory, that escaped notice before November 2009 became causes for worry and a trip to the dealer. Later, the NHTSA examined a number of engine data recorders from accidents where the driver claimed to have experienced acceleration despite applying the brakes. In all cases, the data recorders showed that the brakes were not applied.

In February 2011, the US Department of Transportation announced that a 10-month investigation of the electronic throttle showed no problems.

In April 2011, a jury in Islip, NY took less than an hour to reject a driver's claim that a mispositioned floor mat caused his Toyota to accelerate and crash into a tree. The jury's verdict? Driver error.

As of this writing, we still do not know the actual extent of the problem. But from the evidence to date, it is clear that public awareness of the problem boosted the rate of complaint far out of proportion to its true scope.

Lesson: Your perception of whether you personally experience a problem or benefit is substantially affected by your prior awareness of others' problems/benefits.

Sources: *Wall Street Journal*, July 14, 2010; The Analysis Group (http://www.analysisgroup.com/auto_safety_analysis.aspx—accessed July 14, 2010); *A Today* online, April 2, 2011.

In some situations, we can avoid telling people that they are participating in a study. For example, a marketing study might try different ads or products in various regions without publicizing that they are doing so for research purposes. In other situations, we may not be able to avoid letting subjects know they are being studied, but we may be able to conceal whether they are in the treatment or control group. One way is to impose a dummy treatment on the control group.

Such a dummy treatment is called a *placebo*. It is especially important when we decide how well the treatment worked by asking the subjects. Experience has shown that subjects will often report good results even for dummy treatments. Part of this is that people want to please, or at least not offend, the researcher. Another part is that people may believe in the treatment and therefore think that it helped even when it did not. The researcher may communicate this positive expectation. For this reason, we prefer that neither the subjects nor any researchers in contact with the subjects know whether the subjects are getting the real treatment or the placebo. Then we hope that the researchers will communicate identical expectations to both groups, and the subjects will be equally eager to please or to expect equally good results. Experience has also shown that people respond positively to attention and just being part of a study may cause subjects to improve. This positive response to the knowledge that you are being treated is called the *placebo effect*. More specifically, positive response to the attention of participating in a study is called the *Hawthorne effect*.

We say a study is *single-blind* when the subjects—the hospitals in our medical errors example—do not know whether they are getting the treatment. It is *double-blind* if the staff in contact with the subjects also does not know. It is *triple-blind* if the people who evaluate the results do not know either. These people might be lab technicians who perform lab tests for the subjects but never meet them. They cannot communicate any expectations to the subjects, but they may be biased in favor of the treatment when they perform the tests.

It is not always practical to have all these levels of blinding. A reasonable compromise might be necessary. For our hypothetical study of medical errors, we cannot prevent the hospitals from knowing that they are being studied because we need their agreement to participate. It may be unethical to have the control group do *nothing* to reduce medical errors. What we might be able to do is consult current practices on methods for reducing medical errors and codify them. Then ask the treatment hospitals to implement those best practices PLUS no-fault-reporting and those at the control hospitals to simply implement the basic best practices code. This way, all hospitals receive a treatment but do not know which one is of actual interest to the researcher.

Try It Yourself 1.4

How would you use blinding in a study to assess the effects of watching television on violent behavior?

In addition to the various forms of blinding, we try to keep all other aspects of each subject's environment the same. This usually requires spelling out in great detail what will be done. For example, when no experiment is being conducted, each individual hospital might decide on how to deal with medical errors. In an experiment, we need to agree on a common method that will be applied to all hospitals. We would try to find hospitals that are as similar as possible and maintain patient conditions as similarly as possible. By keeping the two groups the same in every way except the treatment, we can be confident that any differences in the results were due to it. Any difference in the outcome due to nonrandom extraneous factors is called *bias*. Statistical bias is not the same as the type that refers to people's opinions or states of mind.

Try It Yourself 1.5

What factors other than watching television might affect violent behavior? How would you control these in a study to assess the effects of watching television on violent behavior?

Before-After Pairing

We could run our study for a year and measure the total number of medical errors each hospital had by the end of that period. A better strategy is to measure how many errors they had the year before the study as well. Then, we have *paired data*—two measurements on each unit. This allows us to compare the treatment to no treatment *at the same hospitals*. The study reported by the CBC had both before and after data on the same hospital.

Even if we use before/after measurements on the same hospitals, we should also retain the control group. Having both a control group and a treatment group allows us to separate out the improvement due to no-fault-reporting from the improvement due to the more general best practices treatment. Having a control group also controls for trends that affect *all* hospitals. For example, the number of errors could be increasing due to an increased patient load at hospitals, generally, or decreasing due to better doctor training or greater awareness of the issue—perhaps generated by CBC news coverage. The vitamin E study compared two groups over the same time period but did not have before and after data.

Try It Yourself 1.6

How could you use a control group or pairing in a study to assess the effects of watching television on violent behavior?

1.5   WHAT TO MEASURE—CENTRAL LOCATION

Part of the plan for any experiment will be the choice of what to measure to see if the treatment works. This is a good place to review the standard measures with which statisticians are concerned: central location of and variation in the data.

Mean

The *mean* is the average value—the sum of all the values divided by the number of values. It is generally what we use unless we have some reason not to use it.

Consider the following set of numbers: {3 5 1 2}
The mean is $(3 + 5 + 1 + 2)/4 = 11/4 = 2.75$.
You will encounter the following symbols for the mean:
$\bar{x}$ represents the mean of a *sample* from a population. It is written as *x*-bar in inline text.
μ represents the mean of a *population*. The symbol is the Greek letter mu.

Why make the distinction? Information about samples is observed, and information about large populations is often inferred from smaller samples. Statisticians like to keep the two things separate in the symbology.

Median

The *median* is the middle number on a sorted list of the data. Table 1.2 shows the sorted data for both groups in the hospital.

The middle number on each list would be the 13th value (12 above and 12 below). If there is an even number of data values, the middle value is one that is not actually in the data set but rather is the average of the two values that divide the sorted data into upper and lower halves.

TABLE 1.2   Hospital Error Reductions, Treatment, and Control Groups

Control	Treatment
1	2
1	2
1	2
1	2
1	2
1	2
1	2
1	2
1	2
1	2
1	2
1	2
2	2
2	2
2	2
2	2
2	2
2	2
2	3
2	3
3	4
3	4
4	5
4	6
5	9

We find that the median is the same for both lists! It is 2. This is not unusual for data with a lot of repeated values. The median is a blunt instrument for describing such data. From what we have seen so far, the groups seem to be different. The median does not capture that. Looking at the numbers, you can see the problem. In the control group, the numbers coming before the 2 at Position 13 are all ones; for the treatment group they are all 2s. The median reflects what is happening at the center of the sorted data but not what is happening before or after the center.

The median is more typically used for data measured over a broad range where we want to get an idea of the typical case without letting extreme cases skew the results. Let us say we want to look at typical household incomes in the neighborhoods around Lake Washington in Seattle. In comparing the Medina neighborhood to the Windermere neighborhood, using the mean would produce very different results because Bill Gates lives in Medina. If we use the median, it will not matter how rich Bill Gates is—the position of the middle observation will remain the same.

Question 1.3

A student gave seven as the median of the numbers 3, 9, 7, 4, 5. What do you think he or she did wrong?

Mode

The *mode* is the value that appears most often in the data, assuming there is such a value. In most parts of the United States, the mode for religious preference would be Christian. For our data on errors, the mode is 2 for all 50 subjects and 1 for the control group. The mode is the only simple summary statistic for categorical data, and it is widely used for that. At different times in the history of the United States, the mode for the make of new cars sold each year has been Buick, Ford, Chevrolet, and Toyota. The mode is rarely used for measurement data.

Expected Value

The expected value is calculated as follows.

1. Multiply each outcome by its probability of occurring.
2. Sum these values.

For example, suppose that a local charitable organization organizes a game in which contestants purchase the right to spin a giant wheel with 50 equal-sized sections and an indicator that points to a section when the wheel stops spinning. The right to spin the wheel costs $5 per spin. One section is marked $50—that is how much the purchaser wins if the spinner ends up on that section. Five sections are marked $15, 10 sections are marked $5, and the remaining sections are marked $0.

To calculate the expected value of a spin, the outcomes, with the purchase price of the spin subtracted from the prize, are multiplied by their probabilities and then summed.

$$\text{EV} = \left(\frac{1}{50}\right)(\$50 - \$5) + \left(\frac{5}{50}\right)(\$15 - \$5) + \left(\frac{10}{50}\right)(\$5 - \$5) + \left(\frac{34}{50}\right)(\$0 - \$5)$$

$$\text{EV} = -\$1.50$$

The expected value favors the charitable organization, as it probably should. For each ticket you purchase, you can expect to lose, on average, $1.50. Of course, you will not lose exactly $1.50 in any of the above scenarios. Rather, the $1.50 is what you would lose per ticket, on average, if you kept playing this game indefinitely.

The expected value is really a fancier mean; it adds the ideas of future expectations and probability weights. Expected value is a fundamental concept in business valuation and capital budgeting—the expected number of barrels of oil a new well might produce, for example, the expected value of 5 years of profit from new acquisition or the expected cost savings from new patient management software at a clinic.

Percents

Percents are simply proportions multiplied by 100. Percents are often used in reporting as they can be understood and visualized a bit more easily and intuitively than proportions.

Proportions for Binary Data

Definition: **Binary data**

Binary data is data that can take one of only two possible outcomes—win/lose, survive/die, purchase/do not purchase.

When you have binary data, the measure of central tendency is the proportion. An example would be the proportion of the survey approving of the president. The proportion for binary data fully defines the data—once you know the proportion, you know all the values. For example, if you have a sample of 50 zeros and ones, and the proportion for one is 60%, then you know that there are 30 ones and 20 zeros.

For the convenience of software and analysis, binary data are often represented as 0s and 1s. For purely arbitrary reasons, a "1" is called a *success*, but this term has no normative meaning and simply indicates the outcome associated with some action or event of interest. For example, in a data set used to analyze college dropouts, a "1" might be used to indicate dropout. With binary data in which one class is much more scarce than the other (e.g., fraud/no-fraud or dropout/no-dropout), the scarce class is often designated as "1."

1.6   WHAT TO MEASURE—VARIABILITY

If all the hospitals in the control group had one fewer error and all those in the treatment group had two fewer, our job would be easy. We would be very confident that the treatment improved the reduction in the number of errors by exactly one. Instead, we have a lot of variability in both batches of numbers. This just means that they are not all the same.

Variability lies at the heart of statistics: measuring it, reducing it, distinguishing random from real variability, identifying the various sources of real variability, and making decisions in the presence of it.

Just as there are different ways to measure central tendency—mean, median, mode—there are also different ways to measure variability.

Range

The *range* of a batch of numbers is the difference between the largest and smallest number. Referring to Table 1.2, the range for the control group is $5 - 1 = 4$. Note that in statistics the range is a single number.

Try It Yourself 1.7

Referring to the same table, what is the range for the treatment group?

The range is very sensitive to outliers. Recall the two similar Seattle neighborhoods—Windermere and Medina. The range of income in Medina, where Bill Gates lives, will be much larger than the range in Windermere.

Percentiles

One way to get around the sensitivity of the range to outliers is to go in a bit from each end and take the difference from there. For example, we could take the range between the 10th percentile and the 90th percentile. This would eliminate the influence of extreme observations.

Definition: **Pth percentile**

In a population or a sample, the Pth percentile is a value such that at least P percent of the values take on this value or less and at least $(100 - P)$ percent of the values take on this value or more. Sometimes, there is a single value in the data that satisfies this requirement and sometimes there are two. In the latter case, it is best to take the midpoint between the two values that do. Software may have slightly differing approaches that can produce differing answers.

More intuitively: to find the 80th percentile, sort the data. Then, starting with the smallest value, proceed 80% of the way to the largest value.

Interquartile Range

One common approach is to take the difference between the 25th percentile and the 75th percentile.

Definition: **Interquartile range**

The interquartile range (or IQR) is the 75th percentile value minus the 25th percentile value. The 25th percentile is the first quartile, the 50th percentile is the second quartile, also called the *median*, and the 75th percentile is the third quartile. The 25th and 75th percentiles are also called *hinges*.

Here is a simple example: 3, 1, 5, 3, 6, 7, 2, 9. We sort these to get 1, 2, 3, 3, 5, 6, 7, 9. The 25th percentile is at 2.5 and the 75th percentile is at 6.5, so the interquartile range is $6.5 - 2.5 = 4$. Again, software can have slightly differing approaches that yield different answers.

Try It Yourself 1.8

Find the IQR for the control data, the treatment data, and for all 50 observations combined.

Deviations and Residuals

There are also a number of measures of variability based on deviations from some typical value. Such deviations are called *residuals*.

Definition: **Residual**

A residual is the difference between a mean value and an observed value or the difference between a value predicted by a statistical model and an actual observed value.

For 1, 4, 4, the mean is 3 and the median is 4. The deviations from the mean are the differences

$$1 - 3 = -2 \qquad 4 - 3 = 1 \qquad 4 - 3 = 1$$

Mean Absolute Deviation

One way to measure variability is to take some kind of typical value for these residuals. We could take the absolute values of the deviations—{2 1 1} in the above case and then average them: $(2 + 1 + 1)/3 = 1.33$. Taking the deviations themselves, without taking the absolute values, would not tell us much—the negative deviations exactly offset the positive ones. This always happens with the mean.

Variance and Standard Deviation

Another way to deal with the problem of positive residuals offsetting negative ones is by squaring the residuals.

Definition: **Variance for a population**

The *variance* is the mean of the squared residuals, where $\mu =$ population mean, x represents the individual population values, and $N =$ population size.

$$\text{Variance} = \sigma^2 = \frac{\sum (x - \mu)^2}{N}$$

The *standard deviation* σ is the square root of the variance. The symbol σ is the Greek letter *sigma* and commonly denotes the standard deviation.

The appropriate Excel functions are VARP and STDEVP. The P in these functions indicates that the metric is appropriate for use where the data range is the entire population being investigated; that is, the study group is not a sample.

The standard deviation is a fairly universal measure of variability in statistics for two reasons: (i) it measures typical variation in the same units and scale as the original data and (ii) it is mathematically convenient, as squares and square roots can effectively be plugged into more complex formulas. Absolute values encounter problems on that front.

Try It Yourself 1.9

Find the variance and standard deviation of 8, 1, 4, 2, 5 by hand. Is the standard deviation in the ballpark of the residuals, that is, the same order of magnitude?

Variance and Standard Deviation for a Sample

When we look at a sample of data taken from a larger population, we usually want the variance and, especially, the standard deviation—not in their own right but as estimates of these values in the larger populations.

Intuitively, we are tempted to estimate a population metric by using the same metric in the sample. For example, we can estimate the population mean effectively by using the sample mean or the population proportion using the sample proportion.

The same is not true for measures of variability. The range in a sample (particularly a small one) is almost always going to be biased—it will usually be less than the range for the population.

Likewise, the sample variance and standard deviation are *not* the best estimates for the population values because they consistently underestimate the variance and standard deviations in the population being sampled.

However, if you divide by $n-1$ instead of n, the variance and standard deviation from a sample become unbiased estimators of the population values. A mathematical proof is beyond the scope of this course, but you can demonstrate this fact with the "Try It Yourself" exercise below.

Definition: **Sample variance**

$$\text{Sample variance} = s^2 = \frac{\sum (x - \bar{x})^2}{n-1}$$

Definition: **Sample standard deviation**

The sample standard deviation s is the square root of the sample variance.

The appropriate Excel functions are VAR and STDEV.

In statistics, you will encounter the term degrees of freedom. Its exact definition is not needed for this course, but the concept can be illustrated here. Let us say you have three observations and you know that their variance is x. Once you know the first two values, the third is predetermined by the first two and the value for the variance. We say there are $n-1$, in this case, two, degrees of freedom. The denominator in the sample variance formula is the number of degrees of freedom.

Try It Yourself 1.10

In your resampling software, randomly generate a population of 1000 values. It does not matter what population you generate—let us say a population of randomly selected numbers between 0 and 9. In Excel, you can do this with the RANDBETWEEN function. Next, find the variance of this population using the population variance formula. Then, repeatedly take resamples of size 10 and calculate the variance for each resample according to the same population formula. How does the mean of the resample variances compare to the population variance?

Tutorials for this exercise using Resampling Stats for Excel and StatCrunch can be found in the textbook supplements.

For a Box Sampler resampling tutorial based on this exercise, see the file box_sampler_tutorial.pdf.

1.7   WHAT TO MEASURE—DISTANCE (NEARNESS)

The concept of statistical distance is of particular interest to the *data science* community identified at the beginning of the Introduction.

Consider a poll in which respondents are asked to assess their preferences for the musical genres listed below. Ratings are on a scale of 1 (dislike) to 10 (like) and we have poll results from three students (Table 1.3).

TABLE 1.3   Musical Genre Preferences

Person	Rock	Rap	Country	Jazz	New Age
A	7	1	9	1	3
B	4	9	1	3	1
C	9	1	7	2	2

Consider person C. Is she more like person A or person B? Looking at the scores, our guess would be that person C is more like person A. We can measure this distance statistically by subtracting one vector from the another, squaring the differences so they are all positive, summing them so we have a single number, then taking the square root so the original scale is restored.

Definition: **Vector**

A vector is a row of numbers. Vector arithmetic is done by performing the operation on the corresponding elements of each vector, resulting in a new vector of sums, differences, products, and so on.

The statistic that we have described is "Euclidean Distance." Here is the formula, followed by the calculations.

As a general example, assume that we have two vectors, w and x, each containing n values. The Euclidean distance between the two vectors is

$$\text{Euclidean Distance} = \sqrt{(w1 - x1)^2 + (w2 - x2)^2 + (w3 - x3)^2 + \cdots + (wn - xn)^2}$$

If you look carefully at the formula, you might recognize that this is the general formula for the distance between two points from high school geometry. The formula mentioned earlier is for n dimensions, whereas high school math courses usually work with two or three dimensions.

For a specific example, the Euclidean Distance between vectors A and B—a measure of how alike person A is to person B—in the table shown earlier is calculated (Table 1.4).

In the table shown earlier, the sum of the squares of the differences of each row is 145. The square root of 145 is 12.04, which is the Euclidean Distance between the two vectors representing person A and person B. Looking back at the data in Table 1.3, try the following problem:

Question 1.4

Let us say that you own a music store. A, B, and C are all customers of yours, and A and B have both just made purchases. You want to recommend one of these purchases

TABLE 1.4   Euclidean Distance Between A and B

	Rock	Rap	Country	Jazz	New Age
Person A	7	1	9	1	3
Person B	4	9	1	3	1
$(A-B)^2$	9	64	64	4	4
Sum $(A-B)^2$	145				
Euclidean distance	12.04				

for C. Which one would you recommend? See the file euclidean_distance.xls for the answer.

Distance measures are used in statistics for multiple purposes:

- Finding clusters or segments of customers who are like one another.
- Classifying records by assigning them the same class as nearby records.
- Locating outliers, for example, airport security screening.
- Finding the distance to a benchmark. For example, if you have a list of symptoms for an individual, what disease is it closest to?

1.8   TEST STATISTIC

Let us continue with our analysis of the hospital data, using the means for error reduction. Certainly, 2.80 is bigger than 1.88. How much bigger? We could say it is $2.80 - 1.88 = 0.92$ bigger. That is, to say that the treatment seems to reduce the number of errors by nearly one. But there are other ways to look at this. The ratio $2.80/1.88 = 1.49$ gives another comparison. It says the reduction in errors for the treatment group is 1.49 times that in the control group or nearly 50% greater.

Just counting errors treats them all alike. Ideally, we like to have some sort of measurement along a scale. For example, we assign a number of points to indicate the level of severity for each error and we total the points for each hospital. That would require training someone to assign points in a consistent manner from case to case. This could be expensive and wasteful if the person spends most of their time on minor errors that do not actually impact patient health.

As a compromise, our researcher decides to count the number of "major" errors and use that as a measure. The hospitals will be given simple criteria that will allow errors that might be considered major to be identified. Then, a trained expert will make the final decision as to which errors are major. To make sure that there are no differences in how the decision is made from hospital to hospital, the researcher asks that one expert to do all the counting. If possible, relevant records will be submitted anonymously to the expert so that he or she does not know which hospital the records came from or whether that hospital is in the treatment or control group.

Try It Yourself 1.11

In studying the impact of watching television on violent behavior, how would you *measure* television watching and violent behavior in an assessment study? Would you consider an hour of watching Rocky and Bullwinkle reruns to be equivalent to watching an hour of live coverage of the war in Afghanistan or a boxing match? What violence rating would you give to robbing a convenience store, becoming a professional boxer, or joining the army to fight in an active combat theater?

Our test statistic will be calculated as follows.

1. Measure the number of major medical errors for each hospital for the year before and the year after the treatment is initiated and find the reduction: errors before minus errors after.
2. Calculate the mean reduction for the control group and the treatment group.
3. Find the difference: treatment minus control = 0.92.

Important: Throughout this example, we will be talking about "*reductions* in number of errors," not in the number of errors.

A test statistic is the key measurement that we will use to judge the results of the experiment or study.

Test Statistic for This Study

Mean reduction in errors (treatment) minus mean reduction in errors (control).

1.9   THE DATA

After performing the study as planned, the researchers will need to enter the data into a computer and proofread it for errors. After doing that, they obtained the results as shown in Table 1.5.

TABLE 1.5   Reduction in Major Errors in Hospitals (Hypothetical Extension of the Earlier Example)

Row	Hospital#	Treat?	Reduction in Errors
1	239	0	3
2	1126	0	1
3	1161	0	2
4	1293	1	2
5	1462	1	2
6	1486	0	2
7	1698	1	5
8	1710	0	1
9	1807	0	1
10	1936	1	2
11	1965	1	2
12	2021	1	2

TABLE 1.5   (*Continued*)

Row	Hospital#	Treat?	Reduction in Errors
13	2026	0	1
14	202	0	3
15	208	1	4
16	2269	1	2
17	2381	1	2
18	2388	0	1
19	2400	1	2
20	2475	0	4
21	2548	0	1
22	2551	0	2
23	2661	0	1
24	2677	1	4
25	2739	1	2
26	2795	1	3
27	2889	0	5
28	2892	1	9
29	2991	1	2
30	3166	1	2
31	3190	0	1
32	3254	0	4
33	3312	1	2
34	3373	1	2
35	3403	1	3
36	3403	0	1
37	3429	1	2
38	3441	1	6
39	3520	0	1
40	3568	1	2
41	3580	0	2
42	3599	0	2
43	3660	1	2
44	3985	0	2
45	4014	1	2
46	4060	0	1
47	4076	1	2
48	4093	0	1
49	4230	0	2
50	5633	0	2

 Remember, we are counting not the number of errors per hospital but rather the *reduction* in errors.

Database Format

This is a standard database format, which all database programs and most standard-purpose statistical software programs use. The rows represent records or cases—hospitals in this example and the columns represent variables, which are data that change from hospital to hospital. The format has two key features, which is required by most statistical software.

1. Each row contains all the information for one and only one hospital.
2. All data for a given variable are in a single column.

Spreadsheets like Excel can deal with data equally as rows or columns, but statistical software expects rows to be records and columns to be variables.

Technique

Although it is possible to enter data into Excel in the above format, not all statistical analyses in Excel can cope with having "group" as a variable. Some procedures want the observations arranged in columns according to which group they are in.

Let us look at the parts. The first column is simply the row number.

Column 2, hospital, contains *case labels*. These are arbitrary labels for the experimental units—a unique number for each unit. Case labels keep track of the data. For example, if we find a mistake in the data, we would need to know which hospital that came from so we could investigate the cause and correct the mistake. Numerical codes are preferred to more informative labels when we wish to conceal the group to which subjects were assigned.

Column 3 labels observations from the treatment group with a one and those from the control group with a zero.

Column 4 is the number of major medical errors at the year before the study minus the number from the following year. A positive number represents a reduction in medical errors. Note that all the numbers are positive—things got better whether subjects got the treatment or not! This could be due to the Hawthorne effect or any extra care the subjects got from being in the experiment or due to any number of other factors that may have changed between the 2 years.

Relational Database Format

Most statistical procedures work with data that are in the format as mentioned earlier—a single table in which columns have variables and rows have individual records. And most statistics courses considerately provide data for students in this format. However, this is not how most organizations store and use data.

The ability to extract data from relational databases for analysis will be of particular importance to those in the *data science* community identified in the Introduction.

Consider a hypothetical jobs database and start with the following information:

Steve Walters, a scientist with 12 years experience who lives in Palo Alto, CA, has applied for a position at HSBC in London as a Data Scientist; there are two such positions with HSBC. Data munging (the ability to extract data and prepare it for analysis) is a required skill. This information might be presented in a single record as follows:

Candidate	Skill Set	Exp	Home	Company	Location	No. of Positions	Position	Skill
S Walters	Scientist	12	Palo Alto	HSBC	London	2	Data Scientist	Munging

Consider also that James Morgan, a banker with 17 years of experience who lives in NY, is also being considered for a position with HSBC in London, but the position title for his job is Banker and the skill required is fixed equity knowledge. This record might look like this:

Candidate	Skill Set	Exp	Home	Company	Location	No. of Positions	Position	Skill
J Morgan	Banker	17	NY	HSBC	London	2	Banker	Fixed equity

When combined, the database now looks like this:

Candidate	Skill Set	Exp	Home	Company	Location	No. of Positions	Position	Skill
S Walters	Scientist	12	Palo Alto	HSBC	London	2	Data Scientist	Munging
J Morgan	Banker	17	NY	HSBC	London	2	Banker	Fixed equity

This table format is known as a *"flat file."*

Definition: **Flat file**

A flat file is a table that has two dimensions—rows and columns.

Note the redundancy in the columns for company, location, and number of positions. This duplication will be multiplied as we consider more candidates, more companies, and more jobs. In a customer database, for example, hundreds of invoices might be linked to a single customer. It would be nice to have a structure that allowed for a single table of customers (where all their address, demographic, and contact information is stored) and a separate table for invoices, with each invoice linked to a customer by a single customer number. Structured information like this is usually stored not in flat files but in relational databases.

Definition: **Relational database**

A relational database is composed as a set of tables, each of which has a key column used to relate the information in one table to another.

Definition: **Database normalization**

Normalization of a database is the process of organizing data so that it is stored in a set of related tables with defined linkages. Be sure to distinguish this definition of normalization from the statistical term.

For example, the above information might be stored in three separate tables—one for the candidates, one for the employers, and one for the positions (Tables 1.6–1.8).

The left column in each table is a key, used to connect one table to another. Consider the following table, which uses the keys to establish the relationship among these tables (Table 1.9).

We interpret the first row as follows:

Candidate c1, S. Walters (the scientist from Palo Alto with 12 years of experience), has applied for a data scientist job (munging is the required skill) with HSBC in London, which

TABLE 1.6   Candidate Table

Cno	Candidate	Skill Set	Exp	Home
c1	S Walters	Scientist	12	Palo Alto
c2	J Morgan	Banker	17	NY
c3	W Weingart	Graphic designer	17	Berlin
c4	D Hvorostovsky	Baritone	19	London

TABLE 1.7   Employer Table

Eno	Company	No. of Positions	Location
e1	HSBC	2	London
e2	Twitter	10	NY
e3	Royal Opera	3	London
e4	Google	3	Palo Alto

TABLE 1.8   Job Table

Jno	Position	Specialty
J1	Data scientist	Munging
J2	Banker	Fixed equity
J3	Attorney	bankruptcy
J4	Graphic designer	3-D Animation

TABLE 1.9   Relationship Table

Eno	Jno	Cno	Startdate	Rate
e1	j1	c1	November 13, 2013	125
e1	j2	c2	November 25, 2013	220
e3	j3	c4	January 12, 2014	180
etc				

has two positions open. For the particular job he is applying for, the start data is November 13, 2013 and the pay rate is $125,000 per year.

Organizing the data like this reduces duplication but also allows us to query the database in a structured manner and efficiently extract information.

Definition: Structured query language (SQL)

SQL is a programming language used to extract information from relational databases and to manipulate the tables in those databases (e.g., join them together and derive new tables).

Here are a couple of examples of the types of queries supported by SQL:

- List the position descriptions that have been applied for at Twitter.
- List the jobs that S. Walters has applied for.
- How many applications were received in the fourth quarter (Q4) of 2012?

From an analytical perspective, a key feature of SQL is that it can extract data from a relational database and put it into a flat tabular form more amenable to analysis.

An introductory statistics course can do no more than scratch the surface of this topic and show readers one source for data. The small example presented earlier is courtesy of Katya Vasilaky, who teaches a course on SQL and database queries at Statistics.com.

Big Data

Since the turn of the millennium, organizations have found that they have a lot of data already on their hands, or being continuously generated, that yield useful information simply by applying statistical and machine learning models:

- OKCupid, a dating site, uses statistical models with their data to predict what forms of message content are most likely to produce a response.
- Telenor, a Norwegian mobile phone service company, was able to reduce subscriber turnover by 37% by using models to predict which customers were most likely to leave and then lavishing attention on them.
- Allstate, an insurance company, tripled the accuracy of predicting injury liability in auto claims by incorporating more information about vehicle type.

The above examples are from Eric Siegel's *Predictive Analytics* (2013, Wiley).

In other cases, the flow of data can be harnessed for experiments that can be used as the basis for pricing decisions:

- Orbitz, a travel site, has found that it could price hotel options higher for Mac users than for Windows users.
- Staples online store found that it could charge more for staplers if a customer lived far from a Staples store.

The challenge of handling the data, though, is substantial. This challenge is not so much a function of the static size of the data, rather it results from the enormous flow of NEW data.

- Walmart, the large retailer, adds to its database more than 1,000,000 transactions per HOUR.
- JP Morgan reportedly made the decision in 2013 to retain financial data that it had previously been discarding after a set period; the result is that that it must add the equivalent of 2002 Terabyte disks DAILY.

Definition: **Big data**

Big data is a relative term—data are big by reference to the past and to the methods and devices available to deal with them. The challenge big data presents is often characterized by the four Vs—*volume, velocity, variety*, and *veracity. Volume* refers to the amount of data. *Velocity* refers to the flow rate—the speed at which it is being generated. *Variety* refers to the different types of data being generated (money, dates, numbers, text, etc.). *Veracity* refers to the fact that data are being generated by organic-distributed processes (e.g., millions of people signing up for services or free downloads) and not subject to the controls or quality checks that apply to data collected for a study.

For the practitioner of traditional statistical methods, big data introduce a whole level of complexity that was previously absent. A traditional large statistical research study might have involved, say, just 10–15 variables and 5000 records. The data would likely have been collected expressly for the purpose of conducting a study and the scarcity of data, or expense of obtaining it, would most likely have been the main issues.

If you consider the traditional statistical study to be the size of a period at the end of a sentence then the Walmart database is the size of a football field (Figure 1.2).

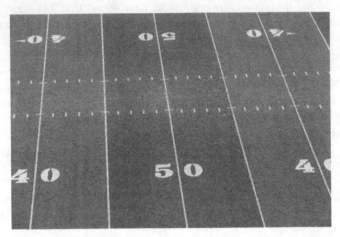

Figure 1.2   In comparing the Walmart database to a traditional statistical study, the difference in scale is like the difference in size between a football field (a real one, not the picture) and the period at the end of this sentence.

Implications for the Practice of Statistics and Statistics Professionals

For traditional research studies involving moderate amounts of data, little has changed about statistical practice or the jobs of statisticians. But the major job growth for statisticians since 2005 has been in the area of what is called *data analytics* or *data science*. Both are somewhat new terms and their definitions are hard to pin down. But central to both is the notion of using statistical and machine learning methods to extract useful information from available organizational data (often of huge size).

The great scale of the flow of new data means that the challenge of extracting, manipulating, cleaning, and preparing data is now enormous, and the time spent doing that easily outweighs the time spent analyzing data. The level of programming expertise required for these steps is substantial. Having gone to great lengths to prepare the data, adding some statistical algorithms into the process to gain interesting knowledge seems like a modest step to the programmer. As a result, statistical models are increasingly finding their way into the repertoire of computer scientists and IT professionals, and statisticians are increasingly called upon to apply their methods to big data.

1.10   VARIABLES AND THEIR FLAVORS

The third and fourth columns in Table 1.5 contain *variables*. These are things we observe, compute, or measure for each subject. They usually vary from one subject to the next. In standard database format, each row represents an experimental unit or subject, while each column represents a variable. Two variables are missing from this table: the number of errors at the beginning of the study and the number at the end. From our point of view, these are intermediate steps.

Quantitative Variables

The numbers in the fourth column are the ones that really interest. They are an example of a *measurement variable* or *quantitative variable*. These are numbers with which you can do meaningful arithmetic. They fall into two types: discrete and continuous.

Definition: **Discrete variable**

The values in a discrete variable differ by fixed amounts and do not assume intermediate values.

The most common type of discrete variable is an integer variable, in which only integers are legal values. Family size is an example. More restricted discrete variables are often the result of rounding or choice of scale. For example, elevations might be any value, but their representations by contour lines on a topographic map are limited to intervals—for example, 50, 100, 150 foot contour lines.

Definition: **Continuous variable**

The values in a continuous variable can assume any values and the difference between any two values can be divided up into any number of legal values.

Age is a continuous variable as is elevation or longitude or latitude. Often, continuous variables may be binned into discrete variables for convenience, as with the contour lines on a map.

Categorical Variables

The other main type of variable is called *categorical* or *qualitative*.

Definition: **Categorical variable**

A categorical variable must take one of a set of defined non-numerical values—yes/no, low/medium/high, mammal/bird/reptile, and so on.

The binary data in column 3 (treatment or control) are categorical variables with two categories. Other examples that might be in the database (although not printed out earlier) are the city, county, or province of each hospital or whether it was a government, business, or charity hospital. Categorical data are often recorded in text labels; for example, male or female, Christian, Muslim, Hindu, Buddhist, Jew, or other. But it is also common to code categories numerically. In our database, treatment is a categorical variable and was coded as one. Control is coded as zero.

Do not do arithmetic on numerical codes for categorical data when it makes no sense. If we code Christian, Muslim, Hindu, Buddhist, Jew, or Other as 1, 2, 3, 4, 5, and 6, respectively, then finding the total or average of these codes is not likely to be meaningful. Coding qualitative data numerically does not make it quantitative! We will see later that some meaningful arithmetic *can* be done on categorical variables.

Computer code in programs will normally have to be told not only what type of variable to expect but also of any limitations on legal values (e.g., that age cannot be negative or that family size must be a positive integer).

Table Formats

Let us see how these data might be presented in other formats. One alternative is to present the error reduction for the control group in one column and the treatment group in the other. This provides the clearest presentation of how the two groups differ in the extent to which errors were reduced. The treatment group had, on average, 2.80 fewer errors in the second year. The control group had 1.88 fewer errors in the second year. Both groups had reduced errors, but the treatment group does appear to have reduced errors to a greater extent than the control group (Table 1.10).

TABLE 1.10   Error Reduction (Hypothetical)

Treatment	Control
2	3
2	1
5	2
2	2
2	1
2	1
4	1
2	3
2	1
2	4
4	1
2	2
3	1
9	5
2	1
2	4
2	1
2	1
3	2
2	2
6	2

TABLE 1.10   (*Continued*)

	Treatment	Control
	2	1
	2	1
	2	2
	2	2
Mean	2.80	1.88

The format in Table 1.10 is one that you might see in print; it might also be used by some softwares—especially Excel.

Technique

Excel's hypothesis testing procedures may require this format in which group is not a separate variable but instead is simply indicated by the column in which a value is located.

You may also see a format that is used only in print, in which all the data are displayed in successive rows for compactness. Neither the rows nor the columns have any significance, so this arrangement is not used with software (Table 1.11).

TABLE 1.11   The Hypothetical Hospital Error Reduction Data in Compact Format for Print

Treatment				
2	2	2	2	2
2	2	2	2	2
4	4	4	4	4
2	2	2	2	2
6	6	6	6	6
Control				
3	3	3	3	3
1	1	1	1	1
1	1	1	1	1
4	4	4	4	4
2	2	2	2	2

1.11 EXAMINING AND DISPLAYING THE DATA

Errors and Outliers are Not the Same Thing!

Suppose that Row 47 of Table 1.5 reads as follows:

Row	Hospital#	Treat?	Reduction in Errors
47	4076	10	2

10 is not a valid value for the "Treat?" variable, which needs to be either zero or one. We would then look up hospital number 4076 to see if we could find the reason for this error. Here is an example of a common error that might lead to a 10 in the Treat? column.

Imagine that we have a study with eight subjects and are supposed to type the eight values 1 0 0 1 0 0 1 0 into a list. Leaving out one space would give 1 0 0 10 0 1 0. As if that 10 were not bad enough, notice that the numbers for subjects six and seven are wrong now and there is no number for subject eight. Noticing this one outlier and trying to correct it helped us to find and fix three other errors.

Definition: **Outliers**

A value (for a given variable) that seems distant from or does not fit in with the other values for that variable is called an *outlier*. It could be an illegal value, as in this case. It could also be a very odd value or a legitimate one. If we saw a 456 in column four, this would not be illegal but it would be a very improbable degree of error reduction.

Some statistical software will identify outliers for you, but keep in mind that these are arbitrary identifications determined by arithmetic. Outliers are not necessarily errors—some are legitimate values. Consider these annual enrollments at a randomly selected set of 10 courses at Statistics.com.

8, 12, 21, 17, 6, 13, 29, 180, 11, 13

The 180 is certainly an outlier, but it is not incorrect. It is the enrollment in an introductory course, whereas the other enrollment figures are for more advanced courses.

Try It Yourself 1.12

Find the average enrollment for the 10 Statistics.com courses whose enrollments are listed above. Would you say this is a good representation of the typical enrollment?

Whenever we find an outlier, we need to investigate it and try to understand the reason for it. If there is an error, we need to try to correct it. Outliers, whether erroneous or legitimate, can strongly affect the numbers we compute from our data. In some cases, an outlier is a symptom of a deeper problem that could have an even greater impact on our results.

Outliers and Social Security

The Social Security Administration is a key source for wage data—Social Security taxes are due on almost all wages and employers must file earnings reports with the tax authorities.

One statistic reported regularly is the average pay of those receiving more than $50 million in wages. This number receives a great deal of attention in the policy debate over income distribution. There were just 74 of these super-earners in 2009 and the government reported on October 15, 2010 that the average income of the super-earners more than quintupled in 2009—to an average of $519 million. This was quite an impressive feat during a severe recession, and the report came during a highly charged political atmosphere in which an important bone of contention was the relative share of income and wealth held by the richest members of society.

Shortly after the report was issued, analysts found that two individuals were responsible for this entire increase. Between them, these two taxpayers reported more than $32 billion in income on multiple W2 (tax) filings. As of November 3, 2010, no information was available on who the individuals were or why they reported such astronomical sums.

However, the Social Security Administration did determine that the filings from the two individuals were in error and issued a revised report. The results?

- 2009 super-earner average wages actually declined by 7.7% from 2008 instead of quintupling.
- 2009 average wages for all workers declined by $598 from 2008; the original report was $384.

These two outliers had a huge and misleading impact on key government statistics. They contributed $214 to the average income of all wage earners, and when they were removed, the recession's wage hit grew by more than 50%. At the same time, the fuel they added to the income distribution debate was illusory.

Question 1.5

What impact would these two outliers have had on statistics that used the median, rather than the mean?

Frequency Tables

Now that we have our results on 50 individual hospitals, we need a way to summarize and compare the treatments and controls as groups. We will look first at summaries that are numbers and then at summaries that are pictures. One numerical summary we could make here is a table of values and how often those values occur, that is, their frequencies (Table 1.12).

This is called a *frequency table* or frequency distribution.

Definition: Frequency table

A frequency table is a table of possible values and the frequencies with which they occur in the data.

Let us interpret a couple of rows.

The bottom row tells us that there were 25 control group hospitals and 25 treatment group hospitals, for a total of 50 hospitals.

The first row tells us that 12 of the 25 control hospitals had a reduction in errors of one and that none of the treatment group hospitals had a reduction in errors of one.

TABLE 1.12   Frequency Distribution—Reduction in Errors

Value	Control	Treatment	Total
1	12	0	12
2	8	18	26
3	2	2	4
4	2	2	4
5	1	1	2
6	0	1	1
9	0	1	1
All	25	25	50

The second row tells us that eight of the 25 control hospitals had a reduction in errors of two and that 18 of the treatment group hospitals had a reduction in errors of two, so that a total of 26 hospitals had a reduction in errors of two.

Notice that if we did not have a control group, we would overestimate the success of the treatment. All the hospitals improved. Still, there is some good news. It looks like the treatment group showed more improvement.

Frequency tables often include *cumulative or relative frequencies*. Here is an example for the control group only (Table 1.13).

TABLE 1.13   Error Reduction Frequency Table (Control)

Error Reduction	Freq.	Cum. Freq.	Rel. Freq.
1	12	12	0.48
2	8	20	0.32
3	2	22	0.08
4	2	24	0.08
5	1	25	0.04
6	0	25	0.00
9	0	25	0.00
All	25	25	1.00

Here is row 2.

$$2 \quad 8 \quad 20 \quad 0.32$$

It illustrates that the value of two for reduction in medical errors showed up eight times in the control group, while a reduction of two *or fewer* errors showed up 20 times and the eight times out of 25 constituted 0.32=32% of the total.

Although it is not shown earlier, we could also calculate *cumulative relative frequency* in the same way that cumulative frequency is calculated—by adding together the current row with the preceding rows. For example, the cumulative relative frequency for the third row (error reduction = 3) is $0.48 + 0.32 + 0.08 = 0.88$.

> ### *Try It Yourself 1.13*
> Compute cumulative and relative frequencies for the treatment hospitals. Can you also find cumulative relative frequencies?

Histograms

We have been looking mostly at single values that summarize the data. Let us return to looking at all the data. Recall the frequency distribution presented earlier and turn it into a picture—a frequency histogram. Figure 1.3 shows a histogram of the error reductions for the treatment group.

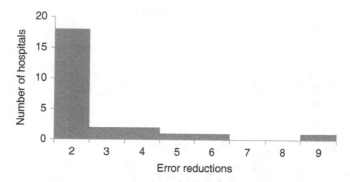

Figure 1.3   Frequency histogram of error reductions in the treatment group.

Interpreting Figure 1.3, we see that 18 hospitals reduced the number of errors by two, two hospitals reduced the number of errors by three, and so on. No hospitals reduced the number of errors by as much as seven or eight, but the histogram *must leave room* for these values to present an accurate picture.

In Figure 1.3, the histogram is relatively easy to make—there are only eight possible values, so we can have a vertical bar for each value.

If we are graphing more complex data—say, hospital sizes—we will not have enough room or visibility to devote one bar to each value. Instead, we group the data into bins. It is important that the bins be (i) equally sized and (ii) *contiguous*. By *contiguous*, we mean that the data range is divided up into equally sized bins, even if some bins have no data, like 7 and 8 as mentioned earlier. Consider Figure 1.4, which shows hypothetical data for hospital sizes in a mid-sized state.

We can see that there were 13 hospitals, with 0–99 beds, 14 hospitals with 100–199 beds, and so on. In Excel's native histogram function, the final bin on the right may include all values larger than a certain amount, but this has the disadvantage of not giving an accurate picture of the gaps in the data.

Deciding on how to display these data is not a trivial matter for a computer. The program must decide on how many values to place in a bin and where the bin boundaries are and the various forms of messiness that can arise. Often, the algorithm that is used results in non-integer values on the *x*-axis, which may not make sense. For example, the binning algorithm in one program produced the following histogram for the hospital error data (Figure 1.5). You can see that it makes much less sense than Figure 1.4.

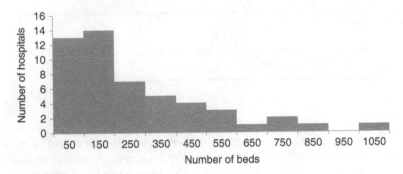

Figure 1.4   Hospital sizes by number of beds (hypothetical data for a mid-sized state).

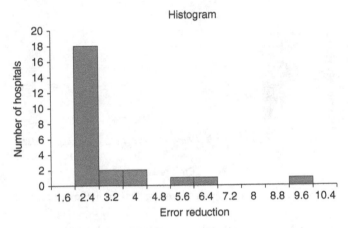

Figure 1.5   Histogram of hospital error reduction, treatment group (*x* values are not integers).

Figure 1.6 shows a *back-to-back* histogram of error reductions. It is described this way because it shows two sets of data—the control set and the treatment set—on the same line, with counts for the control set on the left side of the Errors column and counts for the treatment set on the right side. You may think of it as a horizontal histogram.

Stem and Leaf Plots

A variant of the histogram is the *stem-and-leaf* plot, in which the counts of *x* that you have seen earlier are replaced with numbers denoting the actual values. Figure 1.7 is a stem-and-leaf plot for hospital sizes in terms of number of beds in a hypothetical set of rural counties.

The column to the left of the vertical line indicates the "stem digit," which in this case represents units of 10. The numbers to the right indicate the "leaves," or the unit values. The number of digits on the right—two digits in row 1—tells us that we are counting two hospitals. In row 6, we see five digits, which means five hospitals.

Now, we read across to count the number of beds at each hospital. The first row tells us that there were two hospitals of size 23 [2|33]. The second row tells us there was one 35-bed hospital *and* one 39-bed hospital [3|59]. Again, row 6 tells us there are five hospitals, one with 74 beds, one with 76 beds, and three with 75 beds [7|46555]. The stem-and-leaf plot

```
         Control        Errors        Treatment

    XXXXXXXXXXXX          1

         XXXXXXXX         2    XXXXXXXXXXXXXXXXXX

               XX         3    XX

               XX         4    XX

                X         5    X

                          6    X

                          7

                          8

                          9    X
```

Figure 1.6   Back-to-back histogram.

```
        2 │ 33
        3 │ 59
        4 │ 69
        5 │ 555
        6 │ 179
        7 │ 46,555
        8 │ 23
        9 │ 5
```

Figure 1.7   Stem-and-leaf plot, hypothetical rural hospital sizes.

has the advantage of conveying more information than the histogram, but it is somewhat difficult to manage when the range in the number of digits in the data is larger than 2 or 3. Most software packages do not implement it easily.

Box Plots

Let us look at another graph of the data distribution—one that has features showing the percentiles of the distribution, as well as outliers.

With a boxplot,

- A central box encloses the central half of the data—the top of the box is at the 75th percentile and the bottom of the box is at the 25th percentile.
- The median is marked with a line.
- "Whiskers" extend out from the box in either direction to enclose the rest of the data or most of it. The whiskers stop when they come to the end of the data or when they get further than 1.5 *inter-quartile range* (IQR), from the top and bottom of the box—whichever comes first.
- Outliers beyond the whiskers are indicated with individual markers.

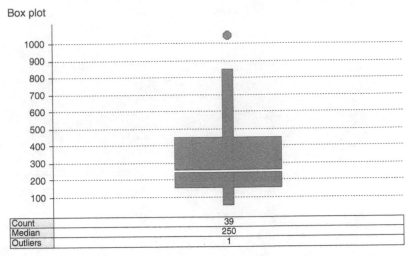

Figure 1.8   Boxplot of metropolitan area hospital sizes (*y*-axis shows the number of beds) (created using Spotfire).

Consider Figure 1.8, which is a boxplot of hospital sizes by number of beds in a hypo-thetical metropolitan area. We can glean the following information.

- Half the hospitals are between 150 beds and 450 beds.
- The IQR is 300 beds.
- The median hospital size is 250 beds.
- The rest of the hospitals are spread out between 50 beds and 850 beds, with the excep-tion of one outlier hospital that has 1050 beds.

Boxplots are a compact way to compare distributions. Below is a side-by-side boxplot comparison of the reduction in errors for the control and treatment hospitals. It was created with XLMiner. Different software has varying ways of creating boxplots. In this case, XLMiner places lines on top of the whiskers, uses a + to indicate the mean, places horizontal V-shaped notches around the mean, and uses o to indicate an outlier (Figure 1.9).

Note how these boxes communicate information by the features that are missing—the top half of the box and the absence of lower whiskers.

Try It Yourself 1.14

What does the absence of the top half of the box and the lower whiskers communicate?

Tails and Skew

Let us review the picture we get from the histograms and the boxplots.

The location of the data is lower for the control group than for the treatment group, which is reflected in the value of the mean.

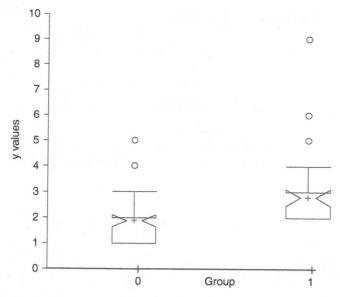

Figure 1.9   Error reductions (*y*-axis) for control hospitals (0) and treatment hospitals (1).

Other than the value of nine, the shape of the distribution for the treatment groups looks roughly like that for the control group. Both groups have peaks at the low end around one or two and trail off toward higher values. We call such a pattern *skewed toward high values*. The part of the picture where the data trail off, say around five, six, seven, eight, or nine, is called the *tail of the distribution*. The direction of the *skew* is the direction of the longer tail. The shape of the distribution is easier to see in the histogram than in the table.

Please look at the spreadsheet ErrorReductions.xls that summarizes all the relevant measures for both the treatment and control data. Be sure that you review and understand the formulas in the highlighted cells.

1.12   ARE WE SURE WE MADE A DIFFERENCE?

We found that the average effect of our treatment was to reduce the number of hospital errors by almost one—0.92. However, we see that the variability from hospital to hospital is more than one. Some people define statistics as the art and science of finding patterns in the midst of variability. We think we found a difference but there is enough variability that it is hard to be sure. For the original study reported by the CBC, we found that those results could well be due to chance. In the next chapter, we will try to determine if the difference we see in this current example is real or might simply be the result of random variability in the numbers.

APPENDIX: HISTORICAL NOTE

Before the widespread availability of computers and software, analysts relied on published tables of random numbers. Here is part of one such table (Table 1.14); the digits are arranged in separate groups of 5 for better visibility.

TABLE 1.14   Portion of a Random Digit Table

58535	99062	55182	89858	67701	94838	37317	10432	75653	78551
56329	09024	81507	90137	19241	55198	74006	52851	41477	58940
04016	38081	45519	27559	92403	30967	86797	17004	22782	09508
37331	94994	67305	34040	91360	83009	36925	31844	12940	51503
24822	53594	72930	23342	88646					

How can we use this table to do the work of 10 coin tosses? We need to convert each random digit into "heads" or "tails." There are several ways to do this. Here are two methods.

Odd = heads, Even = tails.

0–4 = heads, 5–9 = tails.

Let us select a random spot in the table and read off 10 digits. For example, looking at the left center of the table, we see

45,519 27,559

Using 0–4 = heads, 5–9 = tails, this amounts to

HTTHT HTTTT

The result is three heads and seven tails.

1.13   EXERCISES

1. Here are 20 more trials of the exercise you did at the beginning of the readings in which you investigated the model of a random distribution of 10 hospital errors between 2 years, 2008 and 2009. Each row is a trial—10 coin flips. You will use the results to determine whether it is unusual for 10 hospital errors over 2 years to be split 7–3 (2008, 2009), just by chance.

```
Run#
1     HHHTTHTTHH
2     TTHHHTTTHH
3     TTHHTHHTTT
4     HTTHTHHTTT
5     HTHTHTHTTT
6     TTHTTTTHHH
7     HHHTHTHHHH
8     HHTHHHTTTH
9     HHTTTTHTTT
10    THHTTTHTTH
11    TTHHHTTHHT
12    TTTHHHHTHT
13    TTHTHTTTTT
14    THTHHHTTTT
15    THTHHHTTTT
16    HHTTHTHHHH
17    HTTHTHTHTH
18    THHTHHHHHT
19    THHHTTHTTT
20    THTHHHTHTH
```

Each "H" or "T" represents an error. Under our chance model, let us say that "H" means the error happened in 2008 and "T" means 2009. Each row represents one trial (i.e., an allocation of 10 errors).

a. For each of the twenty runs, count the number of times "H" (2008) occurred.

b. Then make a frequency table for your results.

c. What proportion of the runs gave seven or more 2008s?

d. Comment on whether the difference between 2008 and 2009 might have happened by chance.

2. There is controversy over the effectiveness of surgery in treating prostate cancer. Give a design for a study to address this issue. You should address all the issues raised in this week's lesson. For each issue either suggest a way to address that issue or give a reason why there may be no way.

3. Below is a list of numbers of home runs hit by the home runs leaders of the American league during the years 1951–1965:

```
33,52,43,32,37,52,42,42,42,40,61,48,45,49,32
```

Calculate mean, variance, and standard deviation of the data. Note: You can use standard Excel functions or any other software of your choice here. Although this is a small data set, most people would consider that it is the population of interest here—not a sample from some much larger population.

4. Estimate the probability that a family with 10 children would have three or fewer girls. Explain how you arrived at your estimate. Hint: Assume that each successive child has a 50/50 probability of being a male or female, and use dice, coins, or random numbers, not a theoretical formula.

5. Consider the following three customers and the items they have purchased from an online merchant in the last month. A "1" indicates that item was purchased, a "0" indicates that item was not purchased.

	Book	MP3	Power Tool	Tablet	Game
Cust 1	1	0	0	1	1
Cust 2	1	0	1	0	0
Cust 3	0	0	0	1	1

a. Calculate all the possible Euclidean distances between customer pairs.

b. Suppose that you now want to set up a recommender system that alerts a customer when a new item is purchased by a like-minded customer. Which customer would be a good source of recommendations for customer 1?

c. Imagine now that you have millions of items available for purchase, and tens of millions of customers. How could you use Euclidean distance as part of a system to keep a customer from being inundated with recommendations?

Use the PulseNew.xls data for all the remaining questions (see the book website for these and all data sets):

6. Make a frequency table for the variable Ran? Describe what you find briefly in words. Although this can be done by hand, this is a good opportunity to get acquainted with your software program. HINT: In Excel, try the COUNTIF function (make sure Analysis Toolpak and Analysis Toolpak VBA are installed). In Statcrunch, use Stat > Tables > Frequency.

7. The numbers for how many ran and how many did not run seem a bit out of balance. Explain how you would check to see if this could reasonably be considered only due to chance. (Just explain the plan; you do not have to carry it out.) Hint: Rolling dice, tossing coins, or using a random number table might be part of your strategy.

8. Consider a metropolitan area with a diversity of neighborhoods (commercial, shopping, residential, industrial, etc.). Restaurants are not spread evenly throughout the area—they tend to be located in a small number of "restaurant districts." Below are three possible histograms for the distribution of restaurants per neighborhood (number of restaurants is on the x-axis). Which best represents what the above-described distribution of restaurants looks like?

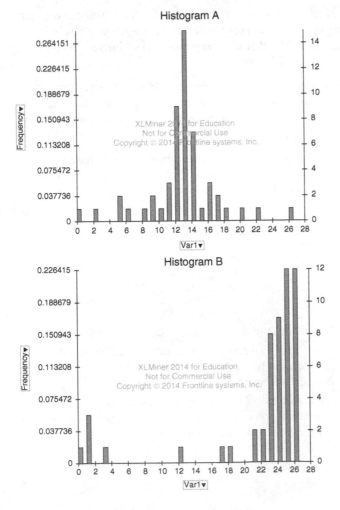

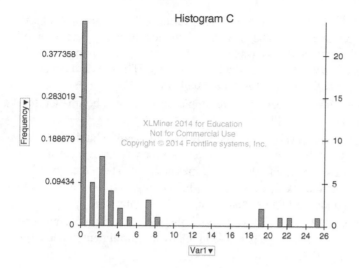

Answers to Try It Yourself

1.1 Results will vary, but chances are this happened at least once.

1.2 Results will vary; in a large class about half the students will get a trial with 14 or more heads, the rest will not.

1.3 You would need to get the subjects to agree to watch TV according to your orders. This might be feasible for a study of short-term effects, but it would be difficult to do if the study required a very long-term period (months or years). Also, some subjects (or their parents) might be averse to watching violent TV, causing dropouts.

1.4 Certainly, the subjects will know what TV shows they are watching. You may be able to conceal that you are studying violence. Those who follow the subjects to measure violent behavior in the future probably can be blind to which subjects received which treatments.

1.5 So many different factors might affect violent behavior that it would be difficult to even list them all, let alone control them.

1.6 A control group could watch no television or watch only nonviolent programs. If we have data for the subjects on other variables we think might also impact violent behavior, then we could pair subjects who are similar on those other variables

1.7 The range is $9 - 2 = 7$.

1.8 Control group IQR: $2 - 1 = 1$
Treatment group IQR: $3 - 2 = 1$
IQR for all observations: $2 - 2 = 0$

1.9

x	mean	Residual	Residual^2
8	4	4	16
1	4	-3	9
4	4	0	0
2	4	-2	4
5	4	1	1

The mean is $(8+1+4+2+5)/5 = 20/5 = 4$. The variance is the average of the sum of the squared residuals $= 6$. The standard deviation is the square root of the variance $= 2.45$. (Note that if this was a sample from a larger population, in calculating the variance, we would have divided by $n - 1 = 4$.) The standard deviation is within the range of the residuals, so it is "in the ballpark."

1.10 If you did enough resamples, the mean of the resample variances will be smaller than the "population" variance.

1.11 Hours spent watching television should be relatively easy to measure but is that what you really want to measure? If we try to determine how much violence each subject watches, we have to find some way to define and measure that. There are a number of websites such as http://www.kids-in-mind.com/help/ratings.htm that offer some rating on a scale of $0-10$, depending on quantity as well as context.

1.12 The average enrollment is 31. This is well above the enrollment for all but one course, so it is not typical. The average is skewed high by the enrollment in one large course—introductory statistics.

1.13 Cumulative, relative, and cumulative relative frequencies

Value	Frequency	Cum. Freq.	Rel. Freq.	Cum. Rel. Freq.
2	18	18	0.72	0.72
3	2	20	0.08	0.8
4	2	22	0.08	0.88
5	1	23	0.04	0.92
6	1	24	0.04	0.96
9	1	25	0.04	1.00
All	25		1.00	

1.14 They indicate that the bulk of the data is "bunched up" at $1-2$ (for the control group) and $2-3$ for the treatment group.

Answers to Questions

1.1 Probabilities must always lie between 0 and 1.

1.2 This would be equivalent to 10 heads in 10 tosses, very unlikely, so chance is not a reasonable explanation.

1.3 The student failed to sort the data before picking the middle value. If the sorting had been done, the numbers would be arranged $3, 4, 5, 7, 9$ so you would see that the median is 5.

1.5 The median would not be greatly affected by these outliers as their presence or absence would hardly affect the value of the middle observation of dozens of observations (for the super-earners) or millions of observations (for the entire wage-earning population).

2

STATISTICAL INFERENCE

The task of trying to assess the impact of random variability on the conclusion drawn from a study, or the results of a measurement, is called *statistical inference*. In this chapter, we look at a particular kind of statistical inference called a *hypothesis test*. Generally, a hypothesis test seeks to determine whether the effects we see in some data from a study are real or might just be the result of chance variation.

Who uses hypothesis testing? The *research* community uses it to determine whether a study is worthy of publication or regulatory approval. *Data scientists* are in less need of the formal apparatus of hypothesis testing but they do use the resampling methods presented here, and their variants, to help separate random from "real" patterns in data.

After completing this chapter, you should be able to

- explain the concept of a null hypothesis,
- describe how to conduct a permutation test with a hat and slips of paper,
- interpret the results of a permutation test,
- describe the shape of the Normal distribution and what is meant when it is said that a more accurate name is the "Error" distribution,
- define, in the context of hypothesis testing, alpha, Type I error, and Type II error,
- explain in what circumstances hypothesis testing is used.

The Null Hypothesis

The standard hypothesis-testing procedure involves a what-if calculation. We ask, "Could my results be due to chance?" This supposition is called the *null hypothesis*. Null is an

Introductory Statistics and Analytics: A Resampling Perspective, First Edition. Peter C. Bruce.
© 2015 John Wiley & Sons, Inc. Published 2015 by John Wiley & Sons, Inc.

old-fashioned word for none or nothing. Here, it means that nothing is really happening, and whatever differences or variations we observe between groups are just due to chance. Then, we calculate how big an effect chance might have in our situation, under the assumption that nothing unusual is going on. If the real-world result that we observe is consistent with the range of outcomes that chance might produce, then we say that what we observed may well be due to chance. In other words, if our observations could be due to chance, we do not reject the null hypothesis. However, if what we observe is much more extreme than what we would expect due to chance, then it is likely that something else is going on.

2.1   REPEATING THE EXPERIMENT

One way to check the results of our experiment would be to repeat it over and over again to see if the result holds. However, doing just one experiment takes a lot of time and money. Repeating it multiple times is out of the question. So instead, we repeat it on a computer, using the model suggested by the null hypothesis or the *null model*.

What if our results were just due to chance? That would mean that the apparent superiority of the hospital treatment group in Chapter 1 was just due to the luck of the draw when we assigned hospitals to treatments. If we assigned them differently, the results would change. Well, let's assign them differently.

Shuffling and Picking Numbers from a Hat or Box

We will take the error reduction scores from the 50 hospitals, put them all in a hat or a box, shuffle the numbers, and then pick them out into two groups of 25. Then, we calculate the mean reduction in errors for each group and the difference in means. In other words, for each pair of groups chosen, we find the mean of the first group minus the mean of the second group.

Try It Yourself 2.1

Stop now and reflect. What is your best guess about the difference in mean error reduction between the two groups of 25 drawn from the hat?

Definition: **Permutation test**

Combining two or more samples in a hat or a box, shuffling the hat, and then picking out *resamples* at random is called a *permutation test*. It can also be done with no hat shuffling by systematically picking out the sample values into two or more resamples in all the different ways possible. This method is sometimes called a *randomization test*, which is an alternate name for a permutation test. If there were two original samples combined in the hat, then you typically draw out two resamples. If there were *n* original samples in the hat, then you would draw out *n* resamples.

> **Hat Video**
>
> The book website contains a video of a permutation test using the hospital error reduction data. Watch the video to see a couple of trials of this permutation test and then continue with the following computer implementation. See the video hat_shuffle-desktop.m4v.

Here is one reshuffling of the medical error reduction scores done by a computer (Table 2.1).

TABLE 2.1   Permutation of Error Reduction Scores into Two Groups

	Random Group 1	Random Group 2
	1	1
	4	2
	1	1
	2	3
	2	1
	3	2
	1	1
	1	2
	2	1
	2	1
	2	4
	5	1
	2	1
	2	6
	2	3
	2	2
	2	9
	4	2
	2	2
	2	3
	2	2
	2	2
	2	5
	2	2
	2	4
Mean	2.16	2.52

The means of the two groups now differ but not by as much as before. Now, they differ only by 0.36. This difference is one estimate of how big a difference might be just due to chance. It would be helpful to have another estimate. In fact, because the computer is doing all the work, why not 50 more? Table 2.2 shows the values of the differences in average error reductions for 50 different permutations.

This is encouraging! We never get a difference as large as 0.92 between the two groups as a result of reshuffling.

TABLE 2.2 Average Error Reduction: First Random Group–Second Random Group, 50 Trials (Continued)

−0.44	0.28
−0.04	−0.60
−0.04	0.04
−0.28	0.04
0.60	−0.44
−0.36	0.60
−0.68	−0.12
−0.12	−0.28
−0.68	−0.60
0.04	0.28
0.12	0.60
−0.76	−0.20
0.04	−0.60
−0.60	0.12
0.12	−0.12
0.44	0.20
−0.28	0.20
0.36	0.60
0.04	−0.04
−0.12	0.04
0.20	−0.12
−0.20	0.28
0.20	0.60
−0.36	−0.20
−0.12	0.44

Try It Yourself 2.2

What do these results mean?

---

p-Value

Earlier, we defined p-value as the probability that the chance model might produce an outcome as extreme as the observed value. In this simulation, the chance model never produced a difference as high as 0.92, so the estimate of the p-value is 0. But we only performed 50 trials. But if we had performed 10,000 trials, we might well see a chance result as high as 0.92, which would raise the estimated p-value slightly above 0.

2.2 HOW MANY RESHUFFLES?

How many shuffles do we need to get good accuracy for the p-value? We just tried 50 and found that the observed result of 0.92 seemed pretty unlikely. Let us now try 1000. Here are the sorted results from the first 20 rows of 1000 trials.

1.08
1
1
1
1
1
1
0.92
0.92
0.92
0.92
0.92
0.92
0.92
0.92
0.92
0.84
0.84
0.84
0.84
+ 980 more.

Only 16 of the 1000 trials had differences in their means of 0.92 or more. Figure 2.1 shows the histogram of all the 1000 trials.

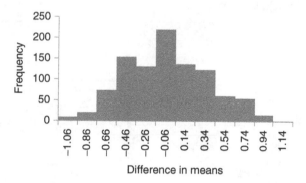

Figure 2.1   Histogram of 1000 trials—permutation of the control and treatment groups (*x*-axis is the mean of the first shuffled group minus the mean of the second shuffled group).

So our impression that the observed result—a difference of 0.92—is not that likely to happen by chance is confirmed. Let us now look at another set of 1000.

1.16
1
0.92
0.92
0.92

0.92
0.92
0.92
0.84
0.84
0.84
0.84
0.84
0.84
0.84
0.84
0.84
0.84
0.84
0.84
+ 980 more.

This time, only 8 of the 1000 trials showed a difference in the means as high as the observed difference—still quite rare.

Let us try it three more times—that is, take three more sets of 1000 trials each.

Boxplots

Now let us compare the overall results of the five sets of 1000 permutations. We can do this with boxplots (Figure 2.2).

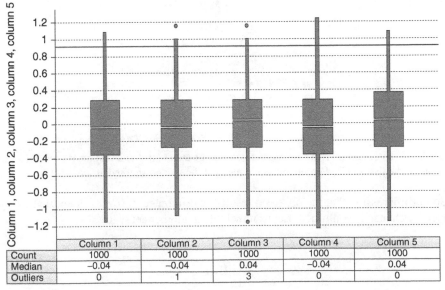

Figure 2.2   Boxplots of five sets of permutation tests, 1000 trials each; y-axis is the difference in the means for simulated pairs of groups, observed difference between means $= 0.92$ (indicated by a bold horizontal line).

Conclusion

Our chance model produced the observed result only rarely, and this was consistent across multiple simulations of 1000 trials each. If you were to count up the extreme (≥ 0.92) results from all the simulations, they would account for less than 2% of the total. This is sufficiently rare to reject the chance model. Two questions, though, may remain in your mind at this point.

1. How rare does rare have to be to reject the chance model? In other words, how low should the p-value be?
2. Is there a firm rule that we can use to determine the number of simulations required?

The answer to question 1 is somewhat arbitrary. If you set the bar for the p-value at a very low value, such as 1%, you will have good protection against being fooled by chance. However, you will also miss more real events because they will appear to be within the range of chance variation. If you set the bar relatively high, such as 10%, more events will qualify as real, but there is a greater probability that some of them are just the result of chance. There is no magic right answer and we will look at this in detail in the following sections.

The great statistician R. A. Fisher is generally regarded as the source of the prevalent standard that a p-value of 5% is the threshold of statistical significance. If the chance model can produce results as extreme as the observed data 5% of the time or more, we cannot rule out chance.

Question 2 also has no right answer. The number of simulations required depends on the purposes of the investigation.

If you are a researcher investigating a new drug, the FDA will require a rigorous presentation of the results. The study will cost tens of millions or perhaps hundreds of millions of dollars. In these circumstances, you will spare no effort or computer time to produce an accurate estimate.

If you are publishing a paper in a scientific research journal, you will be asked to justify how you arrived at your conclusion. Details about the nature of the statistical simulation, the software used, and so on may be required. Again, you will not stint on computer time to produce the most accurate estimate possible.

If you are a data scientist examining some data to assess the possible role of chance, perhaps as a preliminary step before further action, a high level of precision is not required.

As a final step in our example, we performed 20,000 trials. Of these trials, 302 had differences in the means as high as 0.92 for a p-value of $302/20{,}000 = 0.0151$. This confirms the trial results obtained earlier and the conclusion that the chance model is not likely to produce the observed value.

Question 2.1

Alcoholic beverages are legal in the United States, but it is not legal to drive while under the influence of alcohol. In most areas, there are legal limits that define "under the influence" in terms of a certain cut-off for blood alcohol level. What are the pros and cons of setting this cut-off at a high level and at a low level?

The Normal Distribution

At this point, let us use our results to briefly introduce a concept that has been central to statistics for a century. The histogram of our first 1000 reshuffles, as shown in Figure 2.1, looks a lot like a theoretical distribution called the *Normal distribution*. The Normal distribution is symmetric—it has identical tails on both sides. Figure 2.3 shows what a theoretical Normal distribution looks like.

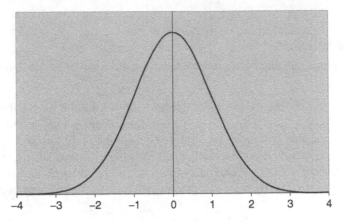

Figure 2.3   The theoretical Normal distribution (the *x*-axis is expressed in standard deviations).

The *x*-axis represents the score of interest and the *y*-axis measures the relative frequency of the scores (more about this later). The *x*-axis shown in Figure 2.3 is expressed in standard deviations. This makes distributions in completely different units and scales comparable to one another.

The Normal Distribution

One misunderstanding about The Normal distribution is that it represents the expected state of nature—that most data are Normally distributed. But this is not the case—most data are *not* Normally distributed.

The Normal distribution was originally called the *error distribution*—it depicted the distribution of not the original data but of *errors* or *deviations* from predicted values. More specifically, it depicted the distribution of errors after the known causes of variability have been removed and all that is left is random noise.

Take heights, for example. At first glance, you might suppose that heights are Normally distributed. But ...

- Women, on average, are shorter than men, so the distribution of heights would have two "humps."
- Children are much shorter than adults.
- People of different countries and of different ethnic groups differ systematically.

If you have removed all those different factors and just considered, say, White women in the United States aged 30–50, then that distribution would more closely resemble the Normal distribution—the remaining variation among individuals would be random noise, which cannot be explained by systematic factors.

Why Is the Normal Distribution Important?

We will use the computer to model probability distributions based on processes such as flipping coins and picking numbers from a hat. You can see that before the advent of the computer, this would have been a major chore.

Use of approximations such as the Normal curve was the standard approach in statistics until the arrival of widely available computing power that allows statisticians to directly model problems with appropriate simulations. Despite the availability of widespread computing power for many decades, most statistical software still use formulas based on Normal and other approximations to implement standard statistical procedures.

The Exact Test

Before we leave this subject, let us look at one more approach. Instead of simply shuffling the error reduction values, we could examine all 126,410,606,437,752 ways of arranging the 50 numbers into two groups of 25. The arrangement of the 126,410,606,437,752 numbers is the complete permutation distribution. Then, we would compare the observed result to this distribution. Obviously, it is going to take some computing power to go that route, but with modern machines and clever algorithms, this is a viable approach in many situations. Hypothesis tests done in this way are known as *exact tests*. Instead of exact tests, it is a lot easier to just try 1000 or 5000 of the possible arrangements, which is what we did. Trying some reasonable subsets of the possible arrangements instead of all of them is how permutation tests are usually carried out.

2.3   HOW ODD IS ODD?

We are now able to conclude that a result as unusual as the result of our experiment—a difference of 0.92—would happen only about 1.5 times in 100 as a result of chance. This equals a *p*-value of 0.0151.

Definition: **Definition repeated from earlier: *p*-value**

The probability of seeing a result from the null model as extreme as the observed value is called the *p-value* or probability value.

Chance seems fairly unlikely as an explanation of our results, so we would reject the null hypothesis and instead conclude that something is going on here. We are pretty sure that the no-fault error reduction program is really effective. This is a different conclusion than the one that we reached for the much more limited data reported by the CBC.

How unlikely does our outcome have to be before we reject the null hypothesis? This is partly for the analyst to decide. Whatever we choose as the answer, the mathematical symbol for it is the Greek letter alpha, α.

Definition: α **(alpha)**

Alpha is the decision threshold you set for the *p*-value in advance of an experiment. For example, you may decide that a *p*-value of less than 0.05, a common choice, lets you rule out chance.

A reasonable value for α depends on the situation. If we are really picky and do not reject the null hypothesis unless α is less than 0.00000001, we will rarely claim a real difference by mistake when there is none—this is called a *Type I error*. On the other hand, we will frequently miss a real effect by not rejecting the null hypothesis. This is called a *Type II error*. In practice, setting a reasonable alpha depends on the relative costs for the two types of errors.

Definition: **Type I error**

When you erroneously conclude that an effect is real when it is just chance, you have committed a Type I error. This occurs when you get a very low p-value, which indicates a low probability of the result happening by chance, but the result is, nonetheless, still due to chance.

Definition: **Type II error**

When you conclude that an effect could be due to chance although it is real, you have committed a Type II error. This occurs when the effect is real, but, due to chance and small sample size, you get a p-value that is not low enough.

Bottom line: When you carry out a study and make decisions about whether the results are interesting and statistically significant, you are balancing the risks and the costs associated with these two types of errors.

For the hospital error example, the social cost of a Type I error would be to implement a program that really does no good. The main financial cost would be the program cost itself. The social cost of a Type II error would be that we discard a helpful program. Then, the cost is measured less in dollars and more in the unnecessary harm due to errors.

For a web-dating service, testing whether a slimmed-down registration form yields more registrants, the cost of a Type I error (and switching to the new form) would be a loss of subscriber data, which is important in creating matches. The cost of a Type II error (sticking with the old form) would be loss of the additional subscribers that would be generated by the new form.

Getting back to the hospitals, if this is just the first stage of research on a program, we would probably be inclined to set the value of alpha moderately high so that if there is even a mild sign it is helpful, we can do further research. Perhaps, an alpha of 10% would be sensible. An alpha of 50% would almost always be too high. This would mean accepting as unusual and acting upon things that happen randomly rather than due to the treatment or other change that we make.

Try It Yourself 2.3

What are the advantages and disadvantages of setting a very high and a very low voting age?

Some researchers advocate setting up a level for alpha in advance. The idea is that you set the rules before playing the game. Then, you cannot decide at the end of the game to let alpha be whatever it needs to be for you to win. While the alpha should be set based on the costs of Type I errors versus Type II errors, many people set it at 5% as some kind of an average figure. This is especially true where a hypothesis test is used as an arbitrary cut-off for narrowing a choice.

An agricultural research station may test 28 varieties of corn in one summer and decide that three are actually better. Then, those might be further developed and tested elsewhere. There is not much cost to either type of error in this case. Even if you have missed one or more good breeds of corn, you have found three, so no one needs to go hungry. If one of the three does not work out in further tests, you have two others, as well as all the varieties from other years. In any case, the costs and benefits are about the same for every variety tested, so it is probably not worth trying to choose a new alpha for every test.

Another type of screening takes place when scientists submit research for publication. In scientific journals, it is very hard to get published with an alpha higher than 5%, so most authors just use that. This has resulted in much criticism, as 5% may or may not be reasonable in any particular situation. Another disadvantage of this policy is that in some situations, it may be worth telling people that something does not work. If an expensive drug is found to provide no improvement over random outcomes, we could save the cost of the drug. Fixing alpha in advance makes the most sense when you know the costs or they do not matter, and you must make a decision based on the outcome of one study.

Another approach does not set the value of alpha in advance but instead lets a reader/user make his or her choice. This approach makes the most sense if you read the studies with an eye to accumulating evidence rather than making a decision. Then, you may not know the costs and may not be paying them anyway, and you would only be interested in the strength of the evidence.

2.4   STATISTICAL AND PRACTICAL SIGNIFICANCE

In the previous section, we were concerned with whether a difference was due to chance or to a treatment. If we decide it is too high to be due to chance, the result is called *statistically significant*. We believe it is real. Knowledge of statistics helps determine statistical significance. This must be contrasted with practical significance—whether the difference is high enough to make a practical difference outside the study in the realm of business, medicine, and so on.

The treatment in our experiment reduced errors by about one per hospital. Whether this is a reduction worth bothering within a medical situation is an issue of medicine and of weighing the costs versus the treatment benefits. The decision is especially difficult if a study shows practical but not statistical significance. This means that we have got a difference that would be important if it were real, but there is enough variability that we cannot conclude that it is real. This is usually the result of a poorly planned or executed study. Perhaps the most common problem is that not enough units were studied. Generally, the ability to detect differences increases with the number of units studied. Statistical power is a measure of our ability to detect a real effect.

When reporting the results of a hypothesis test, we should report both whether the result was statistically significant and just how big it actually was. So for the errors experiment, we would report that we found a reduction of 0.92 in the number of errors as a result of the treatment and that we believe that the difference was not due to chance. A measure of the size of effect is called *effect size*. If possible, a report should include an evaluation by an expert of the practical importance of the size of the effect that we observed.

2.5   WHEN TO USE HYPOTHESIS TESTS

Let us sum up when hypothesis tests are really used:

1. Experiments in business and industry (testing responses to web offers, prototyping a new manufacturing process, trying out a new chemical formulation, etc.). In such a case, the main point is to learn where the experimental results lie, compared to possible chance outcomes (i.e., a *p*-value). Setting alpha is not so important as the decision-makers will want to judge the costs and benefits of implementation, further research, and so on on a case-by-case basis. The *p*-value is a factor, but not necessarily the deciding factor, in such analyses.

2. Studies for publication in scholarly or technical journals. The primary purpose here is to assure the readership that the effect reported is real, not due to chance, so alpha is typically set in advance (usually at 0.05) for a yes/no decision.

3. Studies for submission to governmental authorities (drug development, environmental compliance, and legal cases). Drug development in particular has a long history of government control and a highly structured regulatory regime—an important goal is to assure the public that any claimed effects of drugs are real and not the result of chance outcomes in studies.

The relative importance of formal hypothesis testing in the field of statistics has declined to some extent as companies steadily increase the amount of analysis they do with the vast quantities of data that they generate in the normal course of business. These data are not typically generated as part of a study, and their analysis often does not follow the protocol of an experiment or study that can be assessed with a standard hypothesis test. On the other hand, resampling methods (generating random data, shuffling existing data, sampling with replacement from existing data) are often used in less formal procedures to assess natural variability in data.

2.6   EXERCISES

1. Evidence has been produced that famous people are less likely to die in the month of their birthday than in other months. The (skeptical) hypothesis is that dying is equally likely in any month regardless of birthday. Now, suppose that of 120 celebrity deaths, only seven occurred in the month of their birthday.
   Imagine a hat with 12 cards, each card a month, as well as a list of the 120 celebrity birthdays. We shuffle and pick a card to simulate a death, noting whether it matched the first celebrity birth month. We then repeat this (replacing the card each time, of course), each time noting whether the month picked from the hat matched the next birth month, and so on, until we have gone all the way through the 120 names on the list.
   Then, we repeat this procedure 100 times for a total of 100 simulations. Each time we record the number of matches we got between the 120 picks from the hat (the simulated deaths) and the list of 120 birthdays. We obtained the following frequency distribution. To understand this, look at the second numerical row—it means that 3 of the 100 simulations produced exactly seven matches between the card drawn from the hat and the birthday on the list.
   What is your conclusion and why?

No. of Matches Between Card and Birthday	Frequency
6	1
7	3
8	9
9	20
10	32
11	25
12	7
13	1
14	2

2. In the real world, it is very hard to identify Type I and Type II errors with any certainty as it is hard to know the *true* state of affairs. Each of the following decisions introduces the risk of either a Type I or a Type II error. Indicate which is the possible error (Type I or Type II) and discuss the likely cost of the error.

   a. A new diet pill is introduced to the market after passing an appropriate statistical review by the US Food and Drug Administration.

   b. A web marketer tests two ads against each other. Ad "B" draws a better response than ad "A," so it is used instead of "A."

   c. A manufacturer of silicon chips tests a new process and the preliminary results show that it produces faster chips. However, the improvement is not statistically significant, so the process is not developed further.

   d. A municipal water supplier tests water for the presence of contaminants, and a particular test yields a high level of contaminants. The water supplier issues an alert for residents not to drink or cook with their water.

   e. A parks department tests water in a lake to determine the levels of fecal coliforms. A particular test yields a high level of coliforms, but due to some uncertainty about the reliability of the lab results, no action is taken.

3. A web manager for a retail merchant tests two different presentations of a product to determine whether there is a difference in the length of time that people stay on that product page. He or she records the visit time in seconds per visitor for 50 visitors for each presentation and finds that people viewing presentation A linger 0.33 seconds longer than those viewing presentation B. From the histogram shown below for a permutation test between the two treatments (putting all 100 values in a hat and repeatedly drawing out pairs of random samples of 50 each), is the observed difference within the range of chance variation?

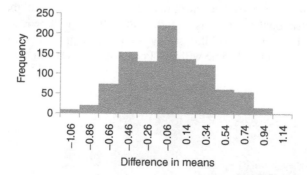

4. From Figure 2.3, the Normal curve, estimate the proportion of values that are
   a. Greater than +2 standard deviations?
   b. Less than −2 standard deviations?

5. Give your assessment of the relative costs of Type I and Type II errors in the following situations where a decision is pending:
   a. A new synthetic feed additive is found in an experiment to accelerate the growth of hogs sufficiently to reduce the "time to market" by 10%. Should it be added to the feed?
   b. Researchers examining longitudinal data that have been collected from a large cohort of nurses notice that those who drink more orange juice have a high incidence of asthma. Should consumption of orange juice be discouraged?
   c. Analysts for a department store conduct an experiment and find that price tags with multiple successive markdowns that are visible on the tag generate more sales than tags with a single markdown indicated, even though the final price was the same in both cases. Should the store switch to a multiple markdown system?
   d. An estate auction manager reviews auction revenue records and finds that auctions conducted in Hyatt hotels generate more revenue than those conducted in Radisson hotels. Should he or she henceforth conduct business only at Hyatt?

Answers to Try It Yourself

2.1 The means are not going to be the same, of course, but there is no reason to expect that the first group will be consistently smaller or larger than the second so the distribution of the mean differences will center around zero. This is a subtle point but an important one that is not obvious to many. If you are unclear about it, try it out with your software. Alternatively, try the following related experiment: Flip a penny 10 times and flip a nickel 10 times. Keep score of "number of penny heads" minus "number of nickel heads" each time. If you had to place a bet on what that score would be the next time, what would you bet on?

2.2 They mean that the observed result is not likely the result of chance. This is a key point—if you are puzzled, stop here and ask questions!

2.3 If you set the voting age too low, irresponsible children will be allowed to vote. If you set it too high, many informed citizens will be denied a voice. No matter what age you set, you will exclude some responsible people and include some irresponsible people. At any age, people vary greatly on their level of responsibility.

Answers to Questions

2.1 If you set the standard for blood alcohol too low, you risk annoying people by arresting them when they are not impaired. If you set it too high, you risk letting impaired drivers on the road. It is important to realize that there is no black and white right answer here in that for almost any level you might set, some drivers will be impaired at that level and some will not. People vary greatly on their susceptibility to alcohol.

3

DISPLAYING AND EXPLORING DATA

So far we have focused on a fairly simple scenario—the number of errors at a hospital. It is not something that requires a graph or a table to understand. Let us now switch to a different scenario—the admission rates of men and women to different departments at a university. We now have three variables—admit/reject, man/woman, and department name. We will look at several simple graphical displays that illustrate what is going on with the data. Then, we will look at time series data and introduce the concept of indexing.

After completing this chapter, you will be able to:

- interpret bar and pie charts,
- determine when to use them,
- convert nominal time series data using an inflation index,
- critique the presentation of time series data with respect to choice of base period,
- properly interpret data that is indexed for relative comparison.

3.1 BAR CHARTS

Let us start with a fragment from a university database; we will also use it later in additional examples (these data are from six UC Berkeley graduate departments for the fall of 1973; they are discussed in Freedman, Pisani, Purves, and Adhikari, *Statistics*, W. W. Norton).

The subjects in this example were applicants to graduate school. The variables are the gender of the applicant, the department to which they applied—which is alphabetically

Introductory Statistics and Analytics: A Resampling Perspective, First Edition. Peter C. Bruce.
© 2015 John Wiley & Sons, Inc. Published 2015 by John Wiley & Sons, Inc.

coded—and whether the applicant was admitted. For this observational study, the data were gathered by simply reading the existing applications.

The most common and useful display for categorical data is the bar chart. It shows bars similar to a histogram with the bar lengths proportional to the number in each category. Bar charts, unlike histograms, are usually drawn with gaps between the bars. Also, unlike histograms, where the *x*-axis shows a progression of values on a continuous scale, the *x*-axis values on bar charts represent categories that are separate from one another and not part of a numerical continuum.

Figure 3.1 is a very crude bar chart for numbers of male and female applicants (total = 8).

The Excel bar chart in Figure 3.2 shows all the data for the total number of applicants to each department.

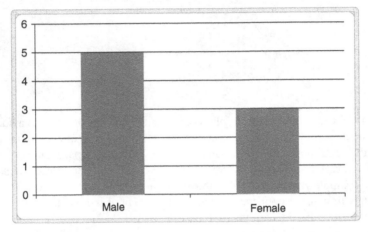

Figure 3.1   Bar chart—eight applicants to a graduate school (excerpt from larger dataset).

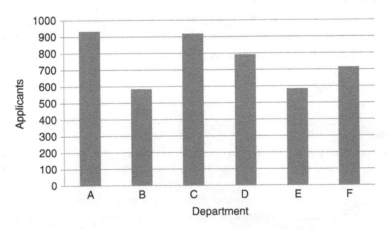

Figure 3.2   Bar chart—applicants by department.

It would have been useful to put the bars in order of length rather than alphabetically. This would make it easier to see that the bar for A is slightly longer than the bar for C, and it would immediately show departments sorted by the number of applications.

Try It Yourself 3.1

Referring to Table 3.1, note that Figure 3.1 is a bar chart for the variable *gender*. Make bar charts (by hand) for the following variables: (1) *department* and for (2) *admit*, showing counts for each value of each variable. Need graph paper? You can find some free at http://statland.org/GraphPaper/gpaper.html.
The first flavor on the list will be fine.

TABLE 3.1   **Applicants to Graduate School (Small Subset)**

Gender	Dept.	Admit
Male	A	Admitted
Male	B	Rejected
Male	A	Admitted
Female	C	Rejected
Male	A	Admitted
Female	B	Rejected
Male	C	Admitted
Female	B	Admitted

The Mode

The mean and median, obviously, cannot be used to describe categorical data. The mode simply names the category that had the largest count, which is the modal category. For the departments in the graduate student study as mentioned earlier, it is A. A modal summary is most useful when the second closest category is far behind unless it is a winner-take-all situation.

3.2 PIE CHARTS

Another common graphical representation of categorical data is the pie chart. It is used when the categories are part of a natural whole. For example, if we had data on all departments that would make a natural whole. In the pie chart shown in Figure 3.3, we could use the artificial whole of all the applicants on whom we have data.

The chart in Figure 3.3 is not very informative. The human eye is much better at comparing lengths than angles, and the pie chart makes it hard to rank the departments by the number of applicants.

Pie charts are most useful when there are few slices of very different sizes, and it is important to visualize the slice compared to the whole. For example, the chart shown in Figure 3.4 shows voting preferences 12 days prior to the November 2, 2010 national elections in the United States. It clearly illustrates the relative balance of the two major parties.

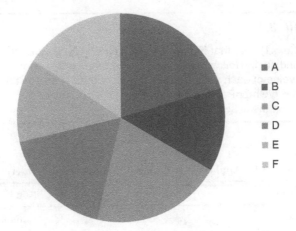

Figure 3.3   Pie chart showing applicants by department.

US voting preference, October 2010

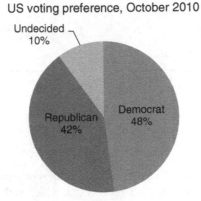

Figure 3.4   Pie chart showing voting preference in US national elections scheduled for November 2, 2010 (Newsweek, http:\\www.newsweek.com, accessed October 24, 2010).

3.3   MISUSE OF GRAPHS

Statistical graphics are a powerful tool for exploring data before crunching any numbers. They are also often misused in the news media or presented in less than optimal ways. Let us look at some issues that come up in the display of graphical information.

Choice of Baseline and Time Period

In 2011, the global economy was in bad shape. Figure 3.5, which is the gross domestic product, shows a dismal picture. GDP was lower at the end of the 6-year period than it was at the beginning.

Recessions are not one-time occurrences. The 1978–1982 period showed similar stagnation, with GDP at the end of the period lower than at the beginning (Figure 3.6).

You may have noted that we chose somewhat arbitrary 6-year and 5-year periods that coincide with the recessions. They are not intended to be representative of the longer term health of the US economy. Observing the 30-year period from 1970 to 2010, you see a much more reassuring long-term trend (Figure 3.7).

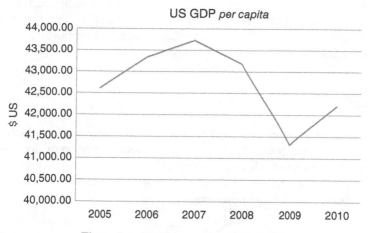

Figure 3.5   GDP *per capita*, 2005–2010.

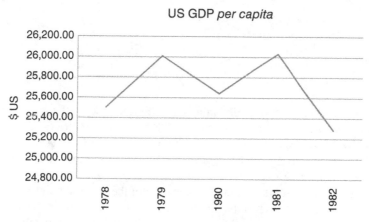

Figure 3.6   GDP *per capita*, 1978–1982.

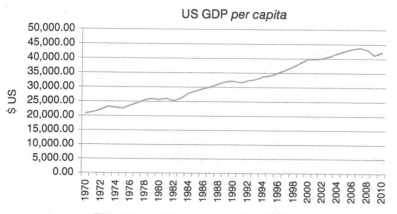

Figure 3.7   GDP *per capita*, 1970–2010.

The lesson to draw from these sets of GDP *per capita* data is that just as small, non-random samples do not paint a reliable picture of the population from which they come, selectively chosen small time periods—particularly when chosen with a particular point in mind—are unreliable in showing long-term trends.

3.4   INDEXING

Perhaps you noticed the term "*per capita*" in the figures shown earlier, which means that the GDP data has been expressed per person. It has also been adjusted for inflation. You may have an intuitive idea of what this means but let us see how it is determined.

Table 3.2 shows a list of figures of US nominal GDP from 2005 to 2010, where "nominal" means not adjusted for inflation or population.

TABLE 3.2   Nominal GDP ($ million)

2005	12,623,000
2006	13,377,200
2007	14,028,700
2008	14,369,100
2009	13,939,000
2010	14,526,500

An increase in income is not really an increase if prices rise by the same amount. So a common adjustment is to divide the nominal GDP by the rise in prices, that is, inflation. If prices rise by 3.24% from 2005 to 2006, then nominal 2006 GDP needs to be adjusted down by 3.24% to express it in constant dollars. We make this adjustment by dividing it by $1 + (3.24/100) = 1.0324$.

The key information here is 2+ places to the right of the decimal point, which obscures it from the typical reader. So when the inflation index is presented, it is typically multiplied by 100 so that the key number (a single-digit inflation rate) is immediately to the left of the decimal and more conspicuous. For example, the adjustment mentioned in the previous paragraph—1.0324—would become 100×1.0324 or 103.24. So, when dividing the nominal GNP by the inflation index, you need to multiply the result by 100. See Table 3.3.

TABLE 3.3   Inflation-Adjusted GDP from 2005 to 2010 ($ million)

	Nominal GDP	Inflation Index	Inflation-Adjusted GDP
2005	12,623,000	100.00	12,623,000
2006	13,377,200	103.24	12,958,500
2007	14,028,700	106.23	13,206,400
2008	14,369,100	108.57	13,234,900
2009	13,939,000	109.73	12,703,100
2010	14,526,500	111.00	13,088,000

For example, to find the inflation-adjusted GDP for 2007, divide the nominal GDP (14,028,700) by the inflation index (106.23) and multiply by 100. (*Note*: The above numbers may not exactly match official sources due to rounding.) This adjustment shows the GDP for each year in 2005 dollars.

Often, you want to know the prior year's GDP in today's dollars. This requires re-expressing prior year GDP levels in terms of 2010 dollars, which means that we want to set the price index for 2010 at 100 and work backwards from there.

Consider again the years 2009 and 2010. The price index rose from 109.73 to 111.00 in 2005 dollars. We now need 2010 to have an index of 100 and the index for 2009 is found by solving for x below:

$$\frac{109.73}{111.00} = \frac{x}{100}$$

$x = 98.86$.

Now, consider the year 2008. We want to find the index for 2008 compared to 2010, and we use the same method.

$$\frac{108.57}{111.00} = \frac{x}{100}$$

$x = 97.81$.

The price indices for 2008 and prior years are found in similar manner.

Inflation-adjusted GDP is then found, as before, by dividing the nominal GDP by the inflation index and then multiplying by 100. See Table 3.4.

TABLE 3.4   GDP Adjustment for Inflation in 2010 Dollars ($ million)

	Nominal GDP	Inflation Index	Inflation-Adjusted GDP
2010	14,526,500	100.00	14,526,500
2009	13,939,000	98.86	14,100,328
2008	14,369,100	97.81	14,690,707
2007	14,028,700	95.70	14,658,625
2006	13,377,200	93.01	14,382,693
2005	12,623,000	90.09	14,011,530

Per Capita Adjustment

For most of us, GDP per person or *per capita* is a more meaningful figure. To find *per capita* GDP, we divide by the population for each year (Table 3.5).

TABLE 3.5   *Per Capita* GDP Adjusted for Inflation ($ million)

	Inflation-Adjusted GDP	Deflator	Population (Thousands)	Inflation-Adjusted *per Capita* GDP
2010	14,526,500	100.00	310,106	46,843.66
2009	14,100,328	98.86	307,483	45,857.26
2008	14,690,707	97.81	304,831	48,192.96
2007	14,658,625	95.70	302,025	48,534.47
2006	14,382,693	93.01	299,052	48,094.29
2005	14,011,530	90.09	296,229	47,299.66

The data for the above calculations can be found in the file gdp.xls. Note that we use a deflator instead of an inflator because we are going backward in years rather than forward. Prices increase as time passes so if we go backward in our calculations, prices go down.

Indexing for Relative Comparisons

Note the role of "100" in the calculations shown earlier. It is used to denote a base or reference level of prices against which other years' prices are compared. In Table 3.4, the price level in 2010 is arbitrarily assigned a value of 100, so all other price levels are expressed relative to the value of 100 in 2010. All prior price levels were lower, of course, due to inflation, so they are all less than 100.

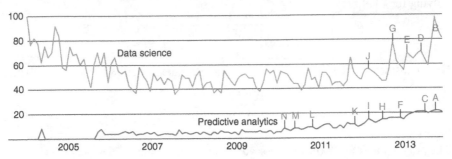

Figure 3.8 Relative number of searches for the terms "data science" and "predictive analytics" indexed at 100. Letters refer to events thought to have a possible impact on traffic; the key to these events is not shown.

A similar approach is used in showing the popularity of search terms on the web. In Figure 3.8, the relative popularity of the search terms "data science" and "predictive analytics" is shown. The underlying data are the number of web searches for each term on a daily basis between 2005 and 2013. The maximum value is noted, and each value is divided by that maximum and then multiplied by 100.

$$y_i = \frac{100 y_j}{y_{\text{max}}}$$

y_j = actual number of searches,

y_{max} = maximum number of searches for either term at any point in the period,

y_i = indexed number of searches.

From this graph you can see that

- "data science" is a more popular search term than "predictive analytics,"
- "data science" had a peak in 2004 and since then declined by half, until 2009, when it started growing again and came back nearly to its prior peak,
- "predictive analytics" has been growing steadily and, as of 2013, is generating about 1/5th the searches that "data science" has been providing.

Google uses the terms "normalization" and "scaling" to refer to this indexing process. The graph shown earlier is based on Google Trends. Graphs at Google Trends add an additional indexing factor—instead of "actual searches" in the above equation, Google uses "searches as a proportion of total searches for all search terms".

For example, to see how the index numbers are calculated, consider a very small hypothetical subset of data, expressed as raw data—number of searches (Table 3.6).

TABLE 3.6   Hypothetical Search Term Data, Number of Searches

	Data Science	Predictive Analytics
January	25,000	11,000
February	23,500	11,700
March	26,100	13,150
April	25,700	9850
May	28,500	14,150

Step 1: Identify the largest value—28,500 for "data science" in May.

Step 2: Divide all values by this number and then multiply by 100 (Table 3.7).

TABLE 3.7   Indexed Hypothetical Search Data

	Data Science	Predictive Analytics
January	88	39
February	82	41
March	92	46
April	90	35
May	100	50

Relative, Not Absolute Comparisons

Now compare the graph on searches for two aircraft types (Figure 3.10) to the graph on searches for the two statistics terms. Which term gets more searches—"predictive analytics" or "Airbus A350?".

At first glance, it looks like "predictive analytics" gets more searches than "Airbus A350." However, the two cannot be compared with each other. The lines for "predictive analytics" and "Airbus A350" represent their performance relative only to "data science" and "Boeing 777," respectively; they do not represent absolute volumes of searches.

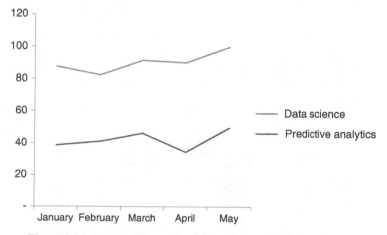

Figure 3.9   Statistics terms: hypothetical indexed search term data.

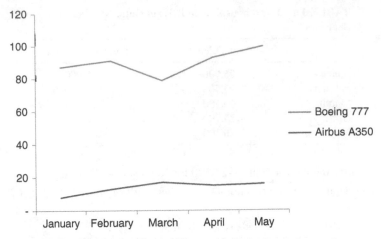

Figure 3.10   Aircraft data: hypothetical indexed search term data.

To learn how "predictive analytics" and "Airbus A350" do relative to each other, you would need to have both search terms on the same plot.

Try It Yourself 3.2

Use Google Search Trends to compare searches for Boeing 777 and Airbus A350. Look at the "Regional Interest" section. Can you figure out how those numbers are calculated?

3.5   EXERCISES

1. In a survey of engineers at a hard drive manufacturer, it was found that 18% were females, 7% were blacks, 35% had degrees in electrical or computer engineering, and 40% were under the age of 35. Would it make sense to present this information in a pie chart? Why or why not?

2. A political commentator makes the following observation in 1991: "From 1973 to 1982, the US economy grew at an annual rate of only 2%. From 1983 to 1990, the growth rate doubled to 4%. This is a big difference." Review the data in GDP.xls and critique this statement, especially with respect to choice of comparison periods.

3. Here is a table of column percents for Department D in the Berkeley study of graduate admissions.

	Female	Male	All
Admitted	34.93	33.09	33.96
Rejected	65.07	66.91	66.04
	100.00	100.00	100.00

Indicate True/False for each of the following statements:

66.91% of male applicants were rejected.

65.07% of those rejected were females.

34.93% of female applicants were admitted.

50% of all applicants were females.

4. Here is a table for Department F at UC Berkeley.

	Admitted	Rejected	All
Female	24	317	341
Male	22	351	373
	46	668	714

Compare the admission rates for men and women in Department F.

5. Consider the following hypothetical data on web searches for the terms "olive oil" and "motor oil":

	Olive Oil	Motor Oil
January	13,000	3000
February	11,000	2000
March	14,000	5000
April	17,000	4000
May	18,000	4000

(a) Convert just the olive oil data into an indexed sequence based on 100 as the maximum value

(b) Convert just the motor oil data into an indexed sequence based on 100 as the maximum value

(c) Convert the two series into two indexed series based on the same index, using 100 as the maximum value

6. (This question requires some easy web searching) Below are listed imaginary revenue figures (in local currency, millions) for telecom companies in several countries. The data are NOT adjusted for inflation.

(a) What is the nominal revenue growth rate for each company?

(b) Which company has the fastest annual growth rate in real terms (i.e., adjusted for inflation in the country where the company is located)?

	2009	2010	2011
Eircom	1500	1650	1815
PLDT	120,000	132,000	145,200
TransTelekom	1100	1210	1331

Note: Eircom is located in Ireland, PLDT in the Philippines, and TransTelecom in Russia.

Answers to Try It Yourself

3.1

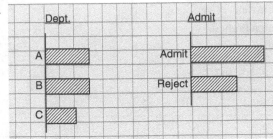

3.2 Results will differ depending on when the search was made. In the "Regional Interest" section of one search for Boeing 777, the first three listings were

United Arab Emirates: 100

Singapore: 83

New Zealand: 83

The algorithm for producing these numbers is probably* as follows:

1. Calculate searches for Boeing 777 as a proportion of all searches for country 1, country 2, and so on. Let p_i = the proportion for the ith country.

2. Find the country with the highest proportion, p_j, set that country = 100.

3. Find the country, k, with the next highest proportion, set that country = x, where

$$x = \frac{p_k * 100}{p_j}$$

4. Continue as above for next highest, and so on.

*Google gives some information online, but not complete details, so this is an educated guess.

4

PROBABILITY

"Is this a game of chance?"

"Not the way I play it."

(W. C. Fields)

Until now, we have been dealing regularly with the notion of probability without really defining it; in this chapter, we introduce more formal concepts of probability. After

Introductory Statistics and Analytics: A Resampling Perspective, First Edition. Peter C. Bruce.
© 2015 John Wiley & Sons, Inc. Published 2015 by John Wiley & Sons, Inc.

completing this chapter, you should be able to:

- produce a Venn diagram,
- use the addition rule and explain in what circumstances it is relevant,
- calculate weighted means and expected values,
- calculate standardized values (z-scores),
- interpret the role of the standard Normal distribution as a benchmark.

What is Probability?

Most people have an intuitive sense of probability and an idea of chance. The weather forecaster does not need to explain what is meant by "chance of rain." Formal and more scientific understandings are more elusive—volumes have been written on probability over the centuries, in the realms of history, philosophy, and mathematics. So, we will not get tangled up in formal definitions of probability. Instead, we will point to two useful concepts in interpreting probability:

1. *Long-run frequency*: Probability can be seen as the frequency with which an outcome would occur of the event, which could be repeated over and over again. For example, the proportion of times you will get "heads" if you flip a coin over and over again. This is easiest to understand for a concrete process, such as a game of chance, whose repetition is easy to visualize. Moreover, in fact, the earliest expositions on the theory of probability were aimed at helping gamblers better understand the odds of games they were playing.

2. *Degree of belief*: Probability can be seen as a numerical mapping of the degree to which you believe something will occur. For example, military planners might attach a probability to the outcome "Pakistan and India will fight a war in the next five years." It is difficult to imagine a repeatable process in which this question can be framed, but the lack of such a process does not diminish the relevance or utility of the concept of probability as applied to the question.

4.1   MENDEL'S PEAS

Gregor Mendel was a nineteenth century Austrian monk who carried out experiments on peas to learn about heredity. At that time, people believed that an offspring exhibited a mixture of the parents' traits. For example, a mule is the offspring of a father, which is a donkey, and a mother, which is a horse, and the mule exhibits characteristics of both.

Mendel's experimental results indicated that things were a bit more complicated. In one experiment, he crossed green peas with yellow peas. On the basis of the mixture theory, one might expect the next generation to be greenish-yellow or some green and some yellow. In fact, all were yellow.

When he crossed peas of this second generation, the next generation of peas was about one-quarter green and three-quarters yellow. Mendel's theory was that each parent contributed something to the color of the offspring. Today, we could call it a gene. In his first cross, each plant had a green gene and a yellow gene. Mendel believed that yellow

Figure 4.1   Gregor Mendel.

beats green genetically, for peas anyway, and we could call yellow a dominant gene. If an offspring gets a yellow gene from either parent, the peas will be yellow. Mendel's first generation of offspring had one pure green parent and one pure yellow parent, so the offspring had both genes. Mendel represented this as Yg, the capitalized Y being the dominant gene. The next generation gets either a Y or a g from each parent, giving the possibilities YY, Yg, gY, and gg. Three out of four peas included a yellow gene, and because yellow is dominant, these peas will all be yellow. The gg peas have no yellow gene and so they will be green.

   YY > yellow
   Yg > yellow
   gY > yellow
   gg > green

Controversy About Mendel's Data

Mendel gathered a huge amount of data in support of his theory. The theory is now accepted, but the data remain controversial. In earlier examples of hypothesis testing, we saw that data are variable and we need to assess how much variability they have before reaching any conclusion. If Mendel had 1000 pea plants in the final generation, we understand that he may not get a perfect 750/250 split. Something close to that would support his theory, while something far away might cause us to reject it.

   The controversy about Mendel's data is that it is too good to be true! It is as if he had performed this testing three times and had obtained 749/251, 750/250, and 751/249. The actual situation is quite a bit more complicated, but one estimate is that one would expect to get that close once in 33,000 tries. How Mendel got to be so lucky is still a subject of controversy.

4.2   SIMPLE PROBABILITY

We can model Mendel's pea experiment with two coin tosses, where heads = Y and tails a g, and we will use this model to introduce some standard probability ideas and terminology. For the second generation of offsprings, we will toss a coin to see if each new offspring gets a Y or a g from the first parent. We will toss a second coin to see what gene they get from the second parent. The possible outcomes are:

Genetically: YY, Yg, gY, and gg
Modeled with a coin: HH, HT, TH, and TT.

For 1000 plants, we expect 500 to get the Y gene from the first parent and the rest to get the g gene. Then, for each of these 500 plants, we expect half to get the Y gene from the second parent. Half of 500 is 250, so on average we expect 250 YY and likewise 250 each of Yg, gY, and gg. The probability of each of these is $250/1000 = 1/4 = 0.25$. This represents the long-term proportion we would expect if we tossed coins or bred peas forever.

Definition: **Sample space**

The list of all possible outcomes of a specified event is called the *sample space*.

The following list shows the probability distribution of all possible outcomes to the event color of the two parent-contributed genes. It lists the sample space and the probability of each item in that space.

YY 0.25
Yg 0.25
gY 0.25
gg 0.25

The sample space must include every possible outcome, and the outcomes on the list must not overlap. We say the outcomes must be jointly exhaustive and mutually exclusive.

Subsets of the sample space are often of interest. The event we might describe in English as "has at least one gene for yellow" would be {YY, Yg, gY}. We will call this event E. The complement of an event is all the outcomes *not* included in the event. So the complement of "has a yellow gene" would be "does not have a yellow gene" or {gg}.

Definition: **An event and its complement**

If we use E to represent an event, we will use $\sim E$ to represent its complement—"not E." For example, Rain and $\sim$Rain—not Rain. Many other notations for complements exist.

The probability rule for events and their complements is

$$P(E) + P(\sim E) = 1$$

Venn Diagrams

It is common to draw pictures called *Venn Diagrams* for simple probability situations.

In Figure 4.2, the dark disk represents event E and the lighter area area represents event $\sim E$. The entire shaded region inside the rectangle, dark + light, represents the whole sample space.

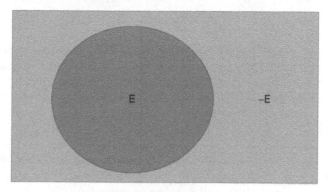

Figure 4.2   Venn diagram.

Another event we might consider is "both genes are the same" or {YY, gg}. We will call this *B*. The event *E* was defined earlier as {YY, Yg, gY}. The *intersection* of *B* and *E*, or any two sets, is the outcomes they have in common. This is represented by $B \cap E$.

In programming logic, the operator *AND* is used: "list all the customers who purchased product *A* and product *B*." For a customer to be listed, he/she must have purchased *both* products.

For the sets *B* and *E*, $B \cap E = \{YY\}$. Figure 4.3 represents two events as the regions bounded by rectangles. Their intersection is the medium-shaded area in common to both E and B.

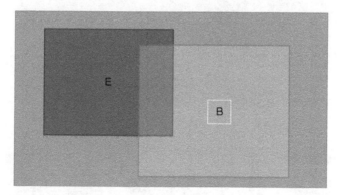

Figure 4.3   Venn diagram—two events of interest and their intersection.

It can happen that two events have no outcomes in common. Then, we represent the outcomes with an empty set, { }, and say that the events are "disjoint" or mutually exclusive. If we know for sure that they are disjoint, the Venn diagram would show no overlaps.

When we computed the probability of $E = \{YY, Yg, gY\}$, we just added up the probabilities for the individual elements to get 0.75. You can do that only if the events are mutually exclusive (disjoint).

Consider the event "both genes are the same" or $C = \{YY, gg\}$ with a probability of 0.5. If we combine C and E together, we cannot get a probability of $0.75 + 0.5 = 1.25$. Probabilities are parts of a whole and can never be more than $1 = 100\%$.

The problem is that YY is in both sets. We cannot count it twice. Looking at the two sets, we have $\{YY, Yg, gY\} \cup \{YY, gg\} = \{YY, Yg, gY, gg\}$, where the U-like symbol is read "union." It means, "dump all the elements together and make a list of what you have without listing anything twice."

In programming logic, the operator *OR* is used: "list all the customers who purchased a tablet or a smartphone." For a customer to be listed, he/she must have purchased *either* product. We list the customers who purchased a tablet, the customers who purchased a smartphone, combine the lists, and remove the duplicates (the ones who purchased both products).

 Do not let your intuition lead you astray with the usage of AND and OR. "AND" may make you think of including all the *outcomes* in the answer – dark, light and medium in Figure 4.3. But we really want to include only those *events that have outcomes B and E together* – that is, just the medium as shown in Figure 4.4. "OR" makes us think of having to choose one outcome or another, but we really want to include all *events* that have *either B or E* as an outcome—dark + light + medium as shown in Figure 4.5.

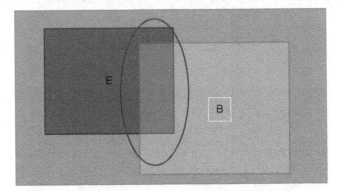

Figure 4.4   "AND" is just the medium-shaded region.

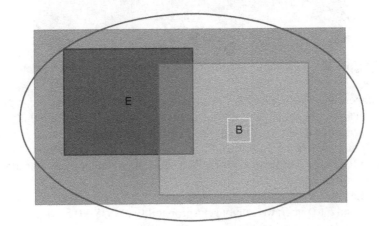

Figure 4.5   "OR" is all three regions.

The formula for the union involves subtracting the intersection (the medium-shaded that is counted in both E and B:

$$P(E \cup B) = P(E) + P(B) - P(E \cap B)$$

Try It Yourself 4.1

Make a table showing the sample space for three tosses of a coin. Assume that the coin has a 50/50 chance of landing heads or tails—that it is a "fair coin." Once you have your table, write out the event "has more heads than tails" as a set and find its probability. Hint: use the table for two tosses as a starting point. Then note that each item on that list has two children. For example, the parent TH has children THT and THH. Hence, there should be eight items on your list instead of four.

The following rule is a simplified version of the above and it applies only in certain circumstances.

ADDITION RULE: P (A or B)

For mutually exclusive (disjoint) outcomes:

$$P(A \cup B) = P(A) + P(B)$$

This reads "the probability of A or B equals the probability of A plus the probability of B." For example, if the probability of the Yankees winning the World Series is 0.2 and the probability of the Nationals winning the World Series is 0.15, then the probability that either the Yankees OR the Nationals will win is 0.35. They cannot both win, so we do not need to subtract out that probability.

4.3 RANDOM VARIABLES AND THEIR PROBABILITY DISTRIBUTIONS

Although there are times when we are interested in an outcome like THHTHTHH, it is more common to be interested in some numerical summary, such as the number of heads, the length of the longest run of heads, or the proportion of heads in the sample. In THHTHTHH, these results are 5, 2, and 0.625, respectively.

Definition: **Variable (noun)**

An attribute (color, size, amount, etc.) that takes on different values from case to case is a *variable* (also called a "feature").

Definition: **Random variable**

A variable that takes on different values (e.g., heads/tails) as the result of a random process is a *random variable*.

The distinction between what is random and what is not is often fuzzy. A coin flip is purely unpredictable. The quantity of acetaminophen in an extra-strength Tylenol is wholly predictable from one tablet to the next—500 mg. However, with some practice, some coins can be flipped so that they always appear to land heads. There is always some tiny variability in the exact amount of acetaminophen per tablet. Most real-life situations have a mix of predictability and chance, but we focus here on the strictly random variable.

Try It Yourself 4.2

For each of your outcomes of tossing a fair coin three times, list the number of heads.

The random variable number of heads in three tosses has possible values of 0–3. We list these and their probabilities in the following probability distribution table.

X	$P(X)$
0	1/8
1	3/8
2	3/8
3	1/8

Verify these probabilities from your counts of the possible outcomes. For example, the event $X = 1$ consists of the outcomes HTT, THT, TTH, or three out of the eight possible outcomes. You may recall that when we did hypothesis tests, we were often interested in the probability that some variable was more than a particular value. From the table shown earlier, you can see that the probability of getting more than one head, that is, two or three heads, is $3/8 + 1/8 = 1/2$. Note that a probability distribution table is always constructed so that the rows are mutually exclusive.

We are also often interested in the mean and standard deviation of a probability distribution. We have already used a shortcut to find the mean or the expected number of heads while tossing a coin. If we toss it n times, we expect $n \times p$ heads. That would be an average of 1.5 heads for our example of three tosses. We will also look at a longer procedure that works for situations other than coin tosses. It is based on the idea of a *weighted mean*.

Weighted Mean

Imagine that you have just finished a course in college and are reviewing your marks:

Final exam: 90%
Hour exams: 87%
Homework: 100%
Quizzes: 100%
Class participation: 20%

Looking at the syllabus that you had got on the first day of class, you note that the instructor had said that the weights attached to various course components were as follows (note that they must add to 100%):

Final exam: 30%

Hour exams: 40%

Homework: 10%

Quizzes: 10%

Class participation: 10%

and in terms of proportions:

Final exam: 0.3

Hour exams: 0.4

Homework: 0.1

Quizzes: 0.1

Class participation: 0.1

How do you compute your final grade? The answer is to multiply each grade by its weight, express it as a proportion of the total grade, and add up the results (Table 4.1).

TABLE 4.1   Calculating Final Grade using Weighted Averages

	Grade (%)	Weight	Product (%)
Final exam	90	0.3	27
Hour exams	87	0.4	35
Homework	100	0.1	10
Quizzes	100	0.1	10
Class participation	20	0.1	2
Overall grade			84

The failure to participate in class was costly. You may remember that in school, one bad grade can pull your average down a lot more than one good grade can pull it up. This is an example of the fact that the means are strongly affected by outliers.

Applying this idea to a probability distribution, we find the mean, or the expected number of heads, by multiplying each possible outcome by its probability and adding up the results. Using this approach, the mean number of heads in three tosses of a coin is shown in Table 4.2.

TABLE 4.2   Expected Number of Heads

Number of Heads	Probability	Product
0	1/8	0
1	3/8	3/8
2	3/8	6/8
3	1/8	3/8
Mean: number of heads		12/8
		= 1.5

Expected Value

We have discussed two equivalent concepts to some extent:

1. Mean (the sum of the values divided by the number of observations).
2. Weighted mean multiply outcomes times their weights, then sum the products.

It might seem simple just to use the mean. Why convert actual data to outcomes and probabilities?

Expected value is used mainly to evaluate the possible future outcomes and their probabilities when there is no simple set of prior data to average:

1. Different possible revenue streams from a new customer of a subscription-based web business.
2. Different possible future valuations of a potential acquisition for a venture capital firm.
3. Different possible profits generated by a new TV series.

Where do the probabilities come from? This is the heart of the problem—they come from the best information that the analyst has available—prior data, your own experience, and others' expertise.

4.4 THE NORMAL DISTRIBUTION

We have mentioned the Normal distribution before and we have pointed out its bell-like shape. Just as not every rectangle is a square, not every bell-shaped curve is a Normal distribution. But the square and the Normal distribution are special enough to merit their own name and some study. The probability situations that we have looked at so far involved discrete situations where we can list and count the possible outcomes. That is, the usual situation for categorical data.

Measurements (continuous variables), on the other hand, can take on any value. The probability that the next patient to walk through Dr. Mason's door will have a weight of 169.82 kg is practically zero. So is the probability that he/she will weigh 55.29 kg. There are so many possibilities that any one of them has next to zero probability.

Any real number is possible for a Normal distribution! It is one of the mysteries of the infinite that under those circumstances, the probability of any particular value is zero. What we are actually likely to be interested is in the probability that the value is between two numbers or is greater or less than some number. In past situations where distributions that looked like Normal distributions came up, we were interested in the probability of getting a certain value or a greater or smaller one. In fact, if the distribution really were a Normal distribution, we could use the Normal distribution to approximate that probability.

With the advent of computers that can toss coins and draw numbers repeatedly from a box, we can calculate these probabilities directly and no longer have to use the Normal distribution to approximate them.

Nonetheless, this distribution is the basis for most of the inference procedures found in textbooks and software and still has important applications. Our goal at this point will be mainly to illustrate the probability rules that we have already studied.

Standardization (Normalization)

Our purpose so far has been to determine how extreme an observed value is relative to a resampling distribution of values under the null hypothesis or chance model. Before computers were available, producing that chance distribution was impractical—it would require lengthy sessions dealing cards from a box or tossing dice.

Instead, as we have noted, approximations to known mathematical distributions were used. It is impractical to derive a different mathematical benchmark each time you want to conduct a study to accommodate the scale of your measurements. A study might have data centered around 15 m with a standard deviation of 3 m and another with a mean of 25 μm with a standard deviation of 2 μm. Instead, we *standardize* our data so that all distributions are on the same scale.

Definition: **Standardization and z-scores**

We standardize or normalize values in a sample or dataset by subtracting the mean from each and then dividing by the standard deviation.

$$\frac{x_i - \mu}{\sigma}$$

Standardized values are also called *z-scores*.

Standardizing data in this way strips scale and units from the measurement. For example, instead of saying that a person weighs 220 pounds, we would say that he weighs 1.8 standard deviations above the mean. You will also encounter the term *normalizing*, which means the same thing. The term *"normalizing"* is not directly related to the Normal distribution—normalizing data does not make it to have a Normal distribution. It does, however, make it easy to compare your data with a standard normal distribution (see below).

The following table illustrates the calculations for cholesterol scores for a group of 10 subjects.

Subject	Cholesterol
1	175
2	210
3	245
4	198
5	210
6	224
7	189
8	171
9	232
10	195
Mean	204.90
Stdev	22.81

Consider subject #6.
Cholesterol raw score: 224

Standardized score: $\frac{224-204.9}{22.81} = 0.837$

So the standardized cholesterol score for subject #6 is 0.837. Another way of saying this is that subject 6's cholesterol is 0.837 standard deviations above the average.

Standard Normal Distribution

The single Normal distribution that is used as a benchmark is called the *standard Normal distribution*.

Definition: **The standard Normal distribution**

The standard Normal distribution is a Normal distribution with a mean of 0 and a standard deviation of 1.

Standard Normal distribution graphs provide standardized values (z-scores) on the x-axis. The y-axis is typically not labeled because it is not very meaningful. What is meaningful is the area *under* the curve. The total area is $= 1$ and the area between or beyond points on the x-axis is the probability that x takes on a value between (or beyond) those points. For example, the shaded area in the graph below is the probability that x is greater than 1 (Figure 4.6).

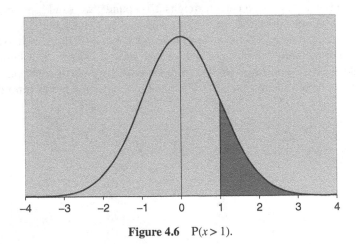

Figure 4.6 $P(x > 1)$.

Calculating this probability from the graph is not easy, so cumulative probability tables are typically used.

Z-Tables

Z-Tables, also called *standard Normal tables*, typically show the probability that a value has a value lower than the specified z-score in the table. Here is a very simple table:

z	$P(Z < z)$
−3	0.001350
−2	0.022750
−1	0.158655
0	0.500000
1	0.841345
2	0.977250
3	0.998650

The first row of numbers in the table shown earlier says that the probability that a value z will be less than −3 is 0.001350 (provided the distribution is standard Normal). So although any number less than −3 is possible, such numbers are not very likely.

The last row of the table says that the probability that Z is less than 3 is 0.998650, which is pretty likely as the cumulative probability must equal 1; we also know that the probability that z is more than 3 is $1 - 0.998650 = 0.001350$. The Normal distribution is symmetrical, so the probability of being more than 3 is the same as the probability of being less than −3.

Older statistics textbooks contain detailed versions of the table shown earlier. It is now more common to find the probability values directly in statistical software or via a web calculator (do a web search for "z-score calculator").

Try It Yourself 4.3

For subject number 8 in the cholesterol table, shown earlier, find (i) the z-score using software or a web-based calculator and (ii) the cumulative probability associated with that score. Then interpret that probability.

The 95% Rule

We can use the table to find the probability that a standard Normal variable is above 2 or below −2 with the rule for unions. The table says that the probability of being below −2 is 0.02275, and we calculated that the probability of being above 2 is also 0.02275. As these are mutually exclusive possibilities, the probability of one or the other happening is additive: $0.02275 + 0.02275 = 0.04550$. Remembering that the standard deviation of the standard Normal is one, and a z of −2 means we are two standard deviations below the mean. A guideline based on the calculation we just performed is that about 95% of any Normal distribution is within two standard deviations of the mean. We used complements again; 4.55% outside means 95.45% inside.

Try It Yourself 4.4

What percentage of the standard Normal distribution is within one standard deviation of the mean? (The guideline is two-thirds.) What about three standard deviations? (The guideline is 99.7%.)

These guidelines are widely used for the detection of *outliers*, which are points unusually distant from the mean and therefore worthy of special attention.

4.5 EXERCISES

1. In basketball, some fouls result in "free throws" (unimpeded shots) by the player fouled. Over his career, a basketball player has scored on 1210 free throw attempts and missed 214 free throw attempts. What is his estimated probability of successfully scoring on a free throw attempt?

2. Consider the following data on the median home value in Boston neighborhoods (from the mid-twentieth century):

```
22
13.1
17.8
20.3
15.4
11.7
25.3
15.2
27.1
23.2
23.1
18.1
32.9
20.3
21.1
21.1
19.9
23.1
16.1
10.4
```

Find the standard Normal score for the first value (22). (For purposes of calculating the standard deviation, you can consider this either as the entire population or as a sample.)

3. Dice are small cubes, held in the hand and then dropped or thrown on a flat surface, as part of a game. Each surface of a die has a number of dots, ranging from one dot to six dots. The number of dots shown on the top surface is said to be how the die lands.
   If you throw a die once, what is the probability that it will land 3? 1 or 6? 1 and 6?

4. Geologists can predict how much oil an oil well will produce but they cannot be certain. Rather, they may express their prediction in probabilistic terms (volume is in barrels per day):

Probability	Volume
0.4	75
0.45	90
0.15	125

What is the expected value of the well's production (in terms of barrels per day)?

5. If you look at the scores on most IQ tests, they have a mean of 100 and a standard deviation of 15.

For class discussion:

(a) Why do you think this is so?

    i. This is the natural state of human intelligence.

    ii. The tests are engineered to make it so for the convenience of comparison and analysis—questions are modified, added, and subtracted over time, so that the mean score stays at 100 and the standard deviation at 15.

(b) Is "IQ" the same thing as "IQ score?" If not, what is the difference?

Homework:

(c) Convert an IQ score of 120 to a z-score.

(d) If you were to select a person at random, what is the probability that they would have an IQ score of 130 or more, assuming the scores are distributed Normally?

Answers to Try It Yourself

4.1

Outcome	Has More Heads?
HHH	Y
HHT	Y
HTH	Y
HTT	N
THH	Y
THT	N
TTH	N
TTT	N

Four of the eight outcomes have more heads than tails = estimated probability of 0.5.

4.2

HHH	Three heads
HHT	Two heads
HTH	Two heads
HTT	One head
THH	Two heads
THT	One head
TTH	One head
TTT	0 heads

4.3 Subject number 8 weighs 171 pounds. The z-score is $\frac{171-204.9}{22.81} = -1.4861$. Using a web calculator*, the cumulative probability is 0.068. This means that the probability of a subject weighing 171 pounds or less is 0.068, *if* the weights are distributed Normally.
*at http://sampson.byu.edu/courses/z2p2z-calculator.html

4.4 The area under the Normal curve outside one standard deviation from the mean (from -1 to $+1$ standard deviations) is $2 \times 0.158655 = 0.317310$, so the area inside -1 to $+1$ standard deviations is $1 - 0.317310 = 0.682690$ (two-thirds is 0.666667). Similarly, the probability of being between -3 and $+3$ standard deviations from the mean is found by subtracting the total area outside the -3 to $+3$ standard deviation range (2×0.001350) from 1, that is, $1 - 2 \times 0.001350 = 0.997300$ which rounds to 99.7%.

5

RELATIONSHIP BETWEEN TWO CATEGORICAL VARIABLES

In this chapter, we look at two-way tables, also called 2×2 tables, in which rows and columns represent binary values of two different variables. 2×2 tables are a subset of $r \times c$ tables (short for row $\times$ column), where the row and columns represent more than two values of their variables. After completing this chapter, you should be able to

- build and interpret 2×2 tables,
- specify how to do a resampling test for a difference between two proportions,
- perform probability calculations involving conditional probabilities,
- perform basic Bayesian calculations
- define and test for statistical independence

5.1 TWO-WAY TABLES

We now return to the data previously mentioned on admission to graduate schools. The data are for the six largest academic departments, and the issue under consideration was admission rates for men and women. We begin with the two categorical variables, Gender and Admit. As before, we look at eight folks in a fragment of the database (Table 5.1).

Ignoring the department variable for now, the first person is a male who was admitted, so he goes in Table 5.2.

Then, we have a rejected male, another admitted male, and a rejected female. We will enter these data as counts in each cell (Table 5.3).

Finishing the table and adding row and column totals gives results that certainly look discriminatory (Table 5.4). However, these are only eight cases out of thousands. Table 5.5 is the full table for all 4526 applicants. Table 5.6 gives the data by percent. The column

Introductory Statistics and Analytics: A Resampling Perspective, First Edition. Peter C. Bruce.
© 2015 John Wiley & Sons, Inc. Published 2015 by John Wiley & Sons, Inc.

TABLE 5.1   Applicants to Graduate School (Small Subset)

Gender	Dept.	Admit
Male	A	Admitted
Male	B	Rejected
Male	A	Admitted
Female	C	Rejected
Male	A	Admitted
Female	B	Rejected
Male	C	Admitted
Female	B	Admitted

TABLE 5.2   Building a 2 × 2 Table

	Female	Male
Admitted		1
Rejected		

TABLE 5.3

	Female	Male
Admitted		2
Rejected	1	1

TABLE 5.4

	Female	Male	Total
Admitted	1	4	5
Rejected	2	1	3
Total	3	5	8

TABLE 5.5   All Applicants

	Female	Male	All
Admitted	557	1198	1755
Rejected	1278	1493	2771
All	1835	2691	4526

TABLE 5.6   All Applicants—Percent by Column

	Female (%)	Male (%)	All (%)
Admitted	30	45	39
Rejected	70	55	61
All	100	100	100

and row labeled "All" are termed *marginal* columns and rows; they give totals by row or column.

The percent table is prepared from the previous one by dividing each entry by the total for the column it is in. These are often called *column percents*. So, for example, 1755/4526 = 38.78% is the overall percentage who were admitted.

Discrimination Against Women?

The percents clarify something that was not so clear in the table of counts: the acceptance rate for males (44.5%) was much higher than the acceptance rate for females (30.3%).

But wait . . . .

Let's look at each department's acceptance rates (Table 5.7).

TABLE 5.7   **Berkeley Admission Rates by Department**

	A (%)	B (%)	C (%)	D (%)	E (%)	F (%)
Female	82	68	34	35	24	7
Male	62	63	37	33	28	6

Female acceptance rates are not that different from male acceptance rates except in Department A, where *males* have a lower acceptance rate. How can females be at such a disadvantage overall?

Simpson's Paradox

The UC Berkeley data are famous as an example of Simpson's Paradox. It is named after Edward Simpson, a British statistician who got his start in statistics at Bletchley Park—the World War II British Decoding Center.

Women have a lower rate of admission than men overall suggesting that there might be discrimination against women. And yet when you look at individual departments, you do not see the apparent discrimination visible in the overall figures as mentioned earlier.

Women had *higher* admission rates in every department except C and E, where their disadvantage was small.

What is going on? How can women have a higher rate of admission in nearly every department but a lower rate overall? Let us look at the actual numbers of applications for each department—originally presented at the beginning of Section 3.1.

Try It Yourself 5.1

Examine the percentages in Table 5.8. Then examine Table 5.9 and where men and women tend to apply. Can you explain the paradox?

TABLE 5.8   **Berkeley Admission Rates by Department**

	A (%)	B (%)	C (%)	D (%)	E (%)	F (%)
Female	82	68	34	35	24	7
Male	62	63	37	33	28	6

TABLE 5.9   Applications to Various Berkeley Department

	A	B	C	D	E	F	All
Female	108	25	593	375	393	341	1835
Male	825	560	325	417	191	373	2691
All	933	585	918	792	584	714	4526

Simpson's Paradox is one of a family of paradoxes or oddities that results from aggregation—putting parts together into a whole. In general, putting parts together results in a whole that looks like the parts, all other things being equal. Simpson's Paradox arises when all other things are not equal.

5.2   COMPARING PROPORTIONS

Let us now look specifically at acceptance rates in Department E (Tables 5.10 and 5.11).

How different is the male acceptance rate from the female acceptance rate? The difference in proportion accepted (female minus male) is $0.2392 - 0.2775 = -0.0383$.

TABLE 5.10   Department E Applicants

	Female	Male	All
Admitted	94	53	147
Rejected	299	138	437
All	393	191	584

TABLE 5.11   Department E Applicants—Percent by Column

	Female	Male	All
Admitted	23.92	27.75	25.17
Rejected	76.08	72.25	74.83
All	100.00	100.00	100.00

Could Chance be Responsible?

Results this close could be due to chance, so we can do a hypothesis test. We ask what things would look like if males and females shared the same overall admission rate of 25.17% or 0.2517. Our null hypothesis is that males and females share the same Admit rate and that the difference between the two groups is due to random allocation.

To solve this problem, we will use software; here is our resampling model.

Note: Box Sampler is limited to a sample size of 200, so cannot be used in this example. In the textbook supplements, you will find the specific procedures for R, Resampling Stats for Excel and StatCrunch.

1. Put 584 chips in a hat to represent the 584 applicants. Of these, 147 are marked admitted and 437 are marked rejected.
2. Shuffle the hat.
3. Draw 393 chips from the hat—the number of females—count the number of admits and calculate this as a proportion.

4. Draw the remaining 191 chips from the hat—the number of males—count the number of admits and calculate this as a proportion.

5. Record the criterion, that is, the statistic of interest. In this case, this statistic is the difference in the acceptance rates, that is, women minus men. The actual difference is $23.92 - 27.75 = -3.83$ percentage points.

6. Repeat the trial many times, perhaps 1000 times, and find out how often we get a difference as extreme as -3.83 percentage points.

The first time we tried this, we obtained -3.57, which makes -3.83 look pretty ordinary. In 1000 trials, 157 differences were more negative than -3.83. Let us round that to a p-value of about 0.16 and say that it is not too unusual. The difference in admission rates between males and females in Department E could just be due to chance.

Put another way, there is no strong evidence of a relationship between gender and admission rate for Department E (Figure 5.1).

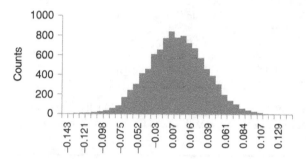

Figure 5.1   Resampling distribution—x-axis is the proportion admit in a group of 393 minus the proportion admit in a group of 191.

Note that once again the distribution shown earlier resembles a Normal distribution. This does not happen always, but it happens often enough to be interesting. A century ago, before we had computers to simulate drawing 584,000 chips out of a hat, that shape was very important. Statisticians worked out a theory that predicted that some of these distributions would be Normal. Then, they could use a table for a Normal distribution to get an approximate p-value without drawing 584,000 real chips out of real hats.

A More Complex Example

The data table on the following page is from the worksheet PulseNew.xls. Some students in a class ran in place, others did not; all measured their pulse at the beginning and at the end of the running activity. Other data are available on each student. We would like to explore whether a relationship exists between two of the variables. For this example, let us choose Smokes? and Sex. To clarify, we ask whether, according to this sample, the data indicate a relationship between smoking and gender.

The full dataset contains 92 subjects in 92 rows of data—the top row contains labels. How can we proceed with our analysis? From the last example, you might decide to create a 2×2 table to illustrate the intersection of smoking and gender. In this case, you would use the variables Smokes? and Sex. A skeleton of the 2×2 table might look like Table 5.12.

TABLE 5.12   Smoking and Gender

Smokes?	Female	Male	All
Yes	?	?	?
No	?	?	?
All (%)	100.00	100.00	100.00

We do not yet know the variable counts for smoking and gender. Please see the file PivotTables.pdf to learn how to create a 2×2 table with these counts using Excel's PivotTable feature. This feature is very useful and you will find it valuable even if Excel is not your primary data analysis tool.

Table 5.13 illustrates the final version of the 2×2 table using percentages.

TABLE 5.13   PivotTable: Smoking and Gender

Smokes?	Sex		Grand Total
	Female (%)	Male (%)	
Yes	22.86	35.09	30.43
No	77.14	64.91	69.57
Grand Total	100.00	100.00	100.00

Can you draw any conclusions about the data? On the basis of this 2×2 table, is it possible to say that females are more likely to be nonsmokers when compared to males?

Try It Yourself 5.2

Using the dataset provided in the PulseNew.xls worksheet, create some 2×2 PivotTables to see if there are any relationships between the categorical variables in the sample. Experiment!

5.3   MORE PROBABILITY

In this section, we use the UC Berkeley data to illustrate the important topic of conditional probability. We also review the probability concepts that we have already studied. We start with the following table of the number of applications.

The gender variable is in the rows and has two categories, while the department variable is in the columns and has six categories. We call this a 2×6 contingency table because it has two rows and six columns and shows all the possible combinations of gender and department.

The cells in this table contain counts, so 593 females applied to Dept. C. The cells must be mutually exclusive and jointly exhaustive; every application has to go in one and only one cell. One person could apply for more than one department, in which case there would be an entry in the table for each application.

Let us pick one application at random from the 4526 applications. Here are some probabilities that we can read from Table 5.14. The probability that a randomly selected

TABLE 5.14   Numbers of Applications to UC Berkeley Departments A, B, C ...

	A	B	C	D	E	F	All
Female	108	25	593	375	393	341	1835
Male	825	560	325	417	191	373	2691
All	933	585	918	792	584	714	4526

application is from a female is 1835/4526. The probability that the application is to Dept. E is 584/4526. The probability that the application satisfies both descriptions—female and Dept. E—is 393/4526.

Try It Yourself 5.3

From the table shown earlier, find the probability that

1. An application is from a female, who applies to dept. A.
2. An application is from a female.
3. An application is to either dept. A or dept. B.
4. An application to dept. B is from a female.
5. An application from a female is to dept. B.

Conditional Probability

Questions 4 and 5 are both conditional probability questions—they ask for the probability of one event, given another event. For example, given that an application is from a female (event one), what is the probability that the application is to department B (event two)?

There were 1835 applications from females and 25 of them applied to department B. This is 1.36% percent or a probability of 0.0136. This probability is "conditioned" on the knowledge that the applicant is female.

Given that an application is to department B, what is the probability that the applicant is female?

There were 585 applications to department B and 25 of them were from females. This is 4.27% or a probability of 0.0427. This probability is "conditioned" on the knowledge that the person has applied to department B.

We implicitly calculate conditional probabilities all the time.

At the time of this writing, the San Francisco Giants were preparing for the 2010 World Series in baseball. They had won 57% of their games in 2010. However, your estimate of their probability of winning their next game would probably be a bit higher if it were conditioned on the knowledge that their ace, Tim Lincecum, would be pitching.

The probability that someone who sees an iTunes promo for a new song then purchases the song might be 0.00001. That probability rises a lot if we know that the person also listens to a demo of the song. The new probability estimate is a conditional probability—the probability of a purchase, given that someone has listened to the demo.

The notation for this relationship of the probability of A, given B, is a vertical line.

P(purchase|demo) = probability of purchase, given demo

In arithmetic terms, what happened is that the denominator in the conditional probability became a lot smaller—it includes only the people who listened to the demo. The formula for a conditional probability is

$$P(A|B) = \frac{P(A \cap B)}{P(B)}$$

In terms of the demo/purchase example, the probability of a purchase, given the demo, is the probability of the purchase AND the demo, divided by the probability of a demo download. Put in terms of numbers, it is the number of people who listened to the demo AND purchased divided by the number of people who listened to the demo. Put in terms of percentages, it is the percentage of demo listeners who purchase.

From Numbers to Percentages to Conditional Probabilities

Table 5.15 takes the original numbers and expresses them as percentages of the overall total.

TABLE 5.15   Percentage of Total Applications by Department and Gender

	A	B	C	D	E	F	All
Female	2.39	0.55	13.1	8.29	8.68	7.53	40.54
Male	18.23	12.37	7.18	9.21	4.22	8.24	59.46
All	20.61	12.93	20.28	17.5	12.9	15.78	100

Such a table is often called a total percents table. It shows each cell as a percentage of the grand total. The 1835/4526 we mentioned earlier turns out to be 40.54%.

We can explicitly present the conditional probabilities in two additional tables. Table 5.16 breaks down each department's applications by gender.

TABLE 5.16   Percentage of Department Applications by Gender

	A	B	C	D	E	F	All
Female	11.58	4.27	64.6	47.35	67.29	47.76	40.54
Male	88.42	95.73	35.4	52.65	32.71	52.24	59.46
All	100	100	100	100	100	100	100

Try It Yourself 5.4

Describe what you see in the table. Do the departments all have about the same ratio of males to females or do they differ quite a bit? If they differ, which departments differ most from what you might expect? Be sure to account for a mostly male applicant pool, which means you would not expect the department ratios to be 50/50.

From Table 5.16, we can say that the probability is 88.42 percent that an application to Department A is from a male. If you go back to the original tables of counts, you can see that this was computed as 825/933. The notation here would be $P(M|A)$—the probability of M given A, or in more explicit English, the probability of picking a male when choosing among the applications to Department A.

Try It Yourself 5.5

From Table 5.16, find $P(M|C)$, $P(\sim M|D)$, and $P(M)$.

The second conditional probability table breaks down each gender's applications by department.

From this, we can see that $P(A|M) = 30.66\%$.

Note that the two conditional probability tables shown earlier break down the applications in different ways, and the values found in one table are nowhere to be found in the other one.

TABLE 5.17   Male/Female Applications by Department (percentage)

	A	B	C	D	E	F	All
Female	5.89	1.36	32.32	20.44	21.42	18.58	100
Male	30.66	20.81	12.08	15.5	7.1	13.86	100
All	20.61	12.93	20.28	17.5	12.9	15.78	100

In plain English, Table 5.16 shows that almost all the applicants to Department A are males, but Table 5.17 shows that less than a third of the males apply to Department A.

 Getting these two probabilities mixed up is a common error in using and interpreting probabilities.

Try It Yourself 5.6

Find $P(E|\sim M)$ and $P(\sim M|E)$. Use the numbers in Table 5.14, and show what counts must be divided to get these numbers.

5.4   FROM CONDITIONAL PROBABILITIES TO BAYESIAN ESTIMATES

Consider a medical screening test that gives a positive result in 98% of the cases where the condition is present but also gives a false-positive result in 3% of cases where the condition is *not* present. Suppose that 0.1% of the people we screen actually have the condition.

Question 5.1

If you test positive, what is the probability that you have the disease?

Let Us Review the Different Probabilities

Shifting now from percentages to probabilities:

1. 0.001 The overall prevalence of the disease is an unconditional probability.
2. 0.98 The probability of a positive test if you have the disease is a conditional probability—P(positive|disease).
3. 0.03 The probability of a positive test if you do not have the disease is a conditional probability—P(positive|no disease).
4. 0.0317 The probability of the disease if you have a positive test is a conditional probability—P(disease|positive).

Probabilities 1–3 are known to the researchers, but probability 4 is of primary interest. It is not known directly—we must calculate it. The best way to see those calculations is visually—stop now to review the Bayes video file at the book web site (or the slides bayes-no-audio.pdf). Figure [Bayesian Calculations] provides an overview.

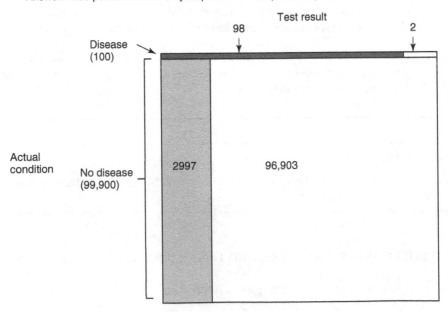

Figure 5.2 Bayesian calculation Venn diagram (medical test example).

Bayesian calculations can be tricky and it is often clearest to work in concrete terms as above. For completeness, the probability formula is shown below, where we know P (B|A) and want to find out P A(|B), and

$$A = \text{disease}, \sim A - \text{no disease}$$

$$B = \text{positive test}$$

$$P(A|B) = \frac{P(A)\,x\,P(B|A)}{P(A)\,x\,P(B|A) + P(\sim B| \sim A)}$$

As an exercise, try translating the above formula into words and relating it to the visual display in the Venn diagram.

Bayesian Calculations

The calculations that we did to determine #4—the probability that you have the disease if you test positive—are termed Bayesian calculations.

The key element is that you have some initial or prior estimate of a probability—the known overall prevalence of the disease. You then receive some pertinent information—the test result—and revise the initial estimate.

In this case, the revision of the initial estimate is surprisingly small and potentially confusing. Faced with a test that is 98% effective in identifying the diseased cases, most people—and many doctors!—have a hard time believing that if you test positive, you have only a 3% chance of having the disease.

While these numbers are made up to simplify calculations, they reflect a real problem for mass screenings in populations where the condition is rare. While you do get true positives for most of the folks having the condition, there are so many more folks who do not have it that their false positives swamp the real positives. This is the reason you often find statisticians testifying against mass screening proposals. To put a human face on it, imagine that folks who test positive will lose their job, be denied insurance, be barred from professional sports, or told they have AIDS, when that is the right decision in only a small percentage of cases.

As this is being written, there is controversy over the use of mammogram screenings in the United States. False-positive rates vary greatly from doctor to doctor and country to country. In the United States, it is estimated that a woman over 40 who has 10 mammograms over a period of 10–20 years has about a 50% chance of obtaining a false positive.

5.5   INDEPENDENCE

In this section, we compare the departments by admission rates (Table 5.18).

> *Try It Yourself 5.7*
>
> What is the probability of being admitted if someone applies to Dept. A? What is the admission probability if someone applies to Dept. F?

TABLE 5.18   Admission Rates by Department

	A	B	C	D	E	F	All
Admitted	64.42	63.25	35.08	33.96	25.17	6.44	38.78
Rejected	35.58	36.75	64.92	66.04	74.83	93.56	61.22
All	100	100	100	100	100	100	100

Is admission rate independent of choice of department?

We would say that admission to a department is independent of choice of department if the admission rate is the same no matter which department you choose. In symbols:

P(Admit|Department) = P(Admit)

Definition: **Independent events**

Two events are independent of one another if the probability that one will occur is unaffected by whether or not the other one occurs.

Whether New York beats Chicago in a baseball game is probably unaffected by whether Los Angeles beats San Diego in a different baseball game—the two events are independent.

On the other hand, whether it rains on Tuesday in Paris is probably connected with whether it rains in Paris on Monday—the two events are connected and not independent.

What about the admission rates? We can see that the admission rate seems independent of gender for departments C and D—an applicant's gender does not greatly alter the probability of admission. This is not true for departments A, B, and F. The variables admit and dept. are independent if the probability of admission is the same for every department. Clearly, they are far from independent here.

Try It Yourself 5.8

This table shows counts for 100 possible outcomes.

	A	~A
B	20	20
~B	40	20

Are A and B independent?

Multiplication Rules

Let us now calculate the probability that both events A *and* B will occur. In general, the probability of A and B is the probability of A times the probability of B, given A.

Multiplication Rule—General Form

$$P(A \cap B) = P(A) * P(B|A)$$

If A and B are independent of one another, knowing that A has happened makes no difference in calculating the probability of B. In this case, P(B) = P(B|A). So when events A and B are independent of one another, the multiplication rule simplifies to this:
Multiplication Rule—Independent Events

$$P(A \cap B) = P(A) * P(B)$$

The probability of A and B is the probability of A times the probability of B.

 The multiplication rule is prone to misuse. The misuse is to blindly assume two events are independent and then use this formula to compute the probability that both happen. For events that are *not* independent, use the more general form of the multiplication rule as shown earlier.

For example, if we take the two baseball games referred to earlier, whose outcomes are independent, let us say we have the following probabilities:
P(New York beats Chicago) = 0.60
P(Los Angeles beats San Diego) = 0.45
Then, the probability that New York wins AND Los Angeles wins = 0.60 * 0.45 = 0.27.

5.6 EXPLORATORY DATA ANALYSIS (EDA)

Until about 1960, statistical graphs and charts were made by hand and carefully drawn in ink. You could neither afford to make very many of them nor could you afford many mistakes. As a result, people tended to proceed directly to calculations. Then, John Tukey invented a whole family of graphics that could be made simply by hand—graphics that did not have to be artistically drawn to tell you a lot about your data. This made it easier to explore your data—to look at it from many different perspectives. The box plots and stem and leaf plots that you have already seen were among Tukey's inventions. Soon after, the traditional displays, as well as Tukey's new displays, were incorporated into computer software. Then, it took no time at all to look at your data from many perspectives.

It is now a standard statistical procedure to explore data from multiple perspectives—both in tabular and graphical form—before proceeding to formulate and answer a detailed/specific question. Statistical methods are almost always applied in some discipline other than statistics. Conclusions are based on knowledge of statistics and knowledge of the subject to which we are applying statistics. When we are lucky, we will have an expert in that area with whom matters could be discussed. Other times, we may just have to say we need to know more.

However, often an initial exploration pays for itself by telling us what information we need. In the UC Berkeley example, there are two questions that need investigation.

1. What determines the departments to which people apply? Had we been able to do an experiment, we could have assigned people to departments. In the observational study that was actually done, people made that choice for themselves. This makes the choice an uncontrolled variable. If that is a free choice, then the apparent discrimination

may be the result of free choices by the applicants rather than a university policy. On the other hand, if females avoid applying to certain departments because they believe those departments treat women poorly, then we may still have a discrimination problem.

2. Why do the acceptance rates vary so much from department to department? This could be the result of supply and demand. When there is a shortage of people in a certain field, academic departments often accept a larger proportion of applicants. On the other hand, women have often complained that traditionally female areas such as nursing are underfunded compared to male-dominated professions. Perhaps, the departments the women are applying to are traditionally female areas that do not have enough funding to educate many students, and so they accept only a few.

Statistics is an ongoing dialog with reality. Sometimes, it gives us answers and sometimes it teaches us what questions to ask.

5.7   EXERCISES

1. An energy company follows the practice of surveying potential natural gas tracts, prior to exercising lease options. The true state of a tract may be positive (economically recoverable natural gas is present) or negative (economically recoverable natural gas is not present). Thirty-five percent of tracts are truly positive. The company's prior experience has been that a positive tract has a 70% chance of yielding a favorable geological survey, but a negative tract still has a 15% chance of yielding a favorable survey. If the company gets a favorable survey on the tract, what is the probability that it is a positive tract?

2. Use the PulseNew.xls data for this question: Make a table to explore the relationship between the variables Ran? and sex. Was the percentage of women who ran higher or lower than that of men?

3. Dice are small cubes, held in the hand and then dropped or thrown on a flat surface, as part of a game. Each surface of a die has a number of dots, ranging from one dot to six dots. The number of dots showing on the top surface is said to be how the die lands.
   (a) If you throw two dice, is the probability of seeing a "1" on the first die independent of the probability of seeing a "1" on the second die?
   (b) If you throw two dice, what is the probability that you will see a "6" on the first die AND a "6" on the second die?
   (c) If you throw two dice, what is the probability that you will see a "6" on the first die OR a "6" on the second die?

4. In 2012, Earnshaw closed 56% of his sales opportunities (14/25), while Samuels closed only 54% (7/13). In 2013, Earnshaw again performed better—71% (10/14) compared with Samuels' 69% (20/29). Yet, over the same two-year period, it turns out that Samuels performed better.
   (a) Calculate the two-year performance for each sales person and confirm Samuels' superior performance.
   (b) Explain the apparent contradiction.

5. A company offers a cloud-sharing service aimed at consumers, including a free level of service to all, as well as higher paying levels. Currently, they offer a relatively low level of free service hoping that people will readily see the need for a higher fee-based level. It decides to run an experiment, offering to some extent a higher level of free service hoping that more people will become engaged users and will upgrade. For several weeks, customers are offered either "low" or "high" levels of service, on a random basis, and then the number of upgrades over the next 3 months is tracked. Here are the results:

Plan	Number of Initial Visitors	Number of Upgrades to "Pay"
Low free service level	329	27
Higher free service level	385	24

   (a) Calculate the difference between the two conversion rates
   (b) Specify a resampling procedure to test whether the difference in the conversion rates might have arisen by chance

6. Consider this hypothetical table on the drug test results and employment status 2 years after the drug test, for a sample of employees:

	Failed Drug Test	Passed Drug Test	Total
Still employed	7	89	96
Not employed	?	?	?
Total	16	?	173

   (a) Fill in the missing information in the table
   (b) Is employment status independent of drug test results?

Answers to Try It Yourself

5.1 The higher overall acceptance rate for men is not due to discrimination against women in the individual departments. Rather, it is due to the fact that men and women have different preferences for departments. More women apply to the tough departments, and this brings down their overall acceptance rate. More men apply to the easy departments, which brings up their overall acceptance rate.

5.3
   1. $108/4526 = 0.0239$ or 2.39%
   2. $1835/4526 = 0.4054$ or 40.54%
   3. $(933 + 585)/4526 = 0.3354$ or 33.54%
   4. $25/585 = 0.0427$ or 4.27%
   5. $25/1835 = 0.0136$ or 1.36%

   Note that we can express the answers either as probabilities (decimals) or percents.

5.4 The applicant pool is about 40% female. The application rates in Departments D and F are somewhat above this, C and E are further above, while A and B have far fewer female applicants than might be expected.

5.5 P(M|C) = 0.354 or 35.4%, which is the probability that the applicant was a male, given that the application was to department C. P(~M|D) = 0.4735 or 47.35%, which is the probability that the applicants were females (not males) given that the person was applying to Department D. P(M) = 0.5946 or 59.46%, which is the probability that an applicant was a male.

5.6 5.6 P(E|~M) = 393/1835 = 0.2142 and P(~M|E) = 393/584 = 0.6729.

5.7 The probability of an application to Department A being accepted is 0.6442, while the probability of an application to Department F being accepted is 0.0644. Note that these differ by a factor of 10!

5.8 Here is the original table with totals and then a column percents table.

	A	~A	All	A	~A	All
B	20	20	40	33	50	40
~B	40	20	60	67	50	60
All	60	40	100	100	100	100

	A	~A	All
B	20	20	40
~B	40	20	60
All	60	40	100

	A	~A	All
B	33	50	40
~B	67	50	60
All	100	100	100

If the events were independent, the three columns would all look the same. As they do not, the events are not independent.

Looking at it another way: The probability of A is different, depending on whether you have B or not-B.

Answers to Questions

5.1 To work out this problem, let us say you screen 100,000 people.

100 of them (0.1%) will actually have the condition, so the other 99,900 do not.

Of the 100 who have the disease, 98 test positive—giving 98 true positives. Two of those with the disease test negative, which is two false negatives.

Of the 99,900 who do not have the disease, 97% test negative, but 3% or 2997 show false-positive results. So we have

- 98 true positives
- 2997 false positives
- 3095 total positives.

Therefore, even if you test positive, the probability is only $98/3095 = 0.0317 = 3.17\%$ that you have the disease (Figure 5.2).

6

SURVEYS AND SAMPLING

We have been discussing surveys and samples; let us now focus a bit on their history and theory. After completing this chapter, you should be able to

- specify what is required for a simple random sample (SRS),
- specify the resampling procedure to determine the sampling distribution of a proportion,
- be conversant with the vocabulary of statistical sampling (samples, populations, parameters, statistics, and sampling frame),
- specify the resampling procedure to determine the sampling distribution of a mean,
- describe and implement the bootstrap,
- describe sampling schemes that may be employed when simple random sampling is infeasible,
- explain the bias caused by self-selection and nonresponse,
- explain the relationship between required sample sizes for a population of 300,000 versus a population of 300 million.

Although survey analysis is considered by some to belong only to the realm of the *research* community, *data scientists* should take note. Big data are not necessarily good data—as we see in the following sections, well-designed small sample surveys can produce more accurate results than huge datasets that are just lying around.

Introductory Statistics and Analytics: A Resampling Perspective, First Edition. Peter C. Bruce.
© 2015 John Wiley & Sons, Inc. Published 2015 by John Wiley & Sons, Inc.

6.1   SIMPLE RANDOM SAMPLES

By the end of 1936, the United States had shown signs of economic recovery from the Great Depression, which started with the collapse of the Wall Street in 1929. GDP was back to where it had been in 1929; it had fallen by a third in the interim. Unemployment headed back to 15%, after having risen to 25% during the depths of the recession. The good news

Figure 6.1   The *Literary Digest* was the premier literary and political commentary weekly of the early twentieth century (image: Wikipedia Commons, LiteraryDigest-19210219.jpg; public domain).

was destined to be short-lived. Another recession hit in 1937, and the economy stagnated until World War II. However, in 1936, things were looking up for President Roosevelt as he campaigned for a second term. He had successfully put Social Security and Unemployment Compensation legislation through Congress and was banking on the popularity of his New Deal platform.

Political polling was in its infancy though, so the preferences of the electorate remained somewhat obscure. In each presidential election year since 1916, the *Literary Digest*, a national weekly opinion magazine that often featured Normal Rockwell illustrations on its cover, had mailed out sample ballots to its readers and used the results to predict the outcome of the vote. The *Literary Digest* poll was a much-anticipated event; up through 1932, it had been accurate. In the summer of 1936, the *Digest* mailed out 10 million ballots and got back 2.4 million. On the strength of the results, the *Digest* predicted a landslide victory for Roosevelt's opponent—Republican Alf Landon.

Republican success in September congressional elections in Maine seemed to validate the Literary Digest's poll. Maine held some nonpresidential elections prior to presidential elections in those days, and was regarded as a bellwether state, inspiring the catch phrase "as goes Maine, so goes the nation (Figure 6.1)."

As it turned out, Roosevelt won in a landslide, capturing 62% of the popular vote and winning every state except Maine and Vermont. Maine's catch phrase was amended and it became "as goes Maine, so goes Vermont." The *Literary Digest* went out of business shortly after the election.

What happened?

In addition to polling its readers, the *Literary Digest* also mailed ballots to lists of automobile owners and telephone subscribers. At a time when a majority of the nation was jobless and destitute, those who could afford magazine subscriptions, automobiles, and telephones were hardly representative of the population. They were wealthier and more Republican than the average voter, and therefore produced a biased prediction.

Definition: **Bias**

A statistical procedure or measure is statistically biased if, when you apply it to a sample from a population, it consistently produces over-estimates or under-estimates of a characteristic of that population.

In 1935, the year before the *Literary Digest* polled its readers, George Gallup, a young advertising executive with Young and Rubicam, founded the American Institute of Public Opinion. It was dedicated to the measurement of public opinion via statistically designed surveys. He was convinced that what mattered was not the quantity of people surveyed but rather their representativeness—the degree to which they reflected the views of the general voting population.

The Gallup Poll

In October, 1935, George Gallup published the first Gallup Poll—America Speaks. The first question was, "Do you think expenditures by the government for relief or recovery are too little, too great, or just about right?"

Sixty percent said "too great."

Gallup quickly capitalized on the success of his political polling (see below). In the 1940s, he teamed up with David Ogilvy, famous for the phrase, "Advertising is neither entertainment nor an art form—it is a medium of information." Together, they worked with Hollywood executives to develop methods for predicting the box-office revenues of films, based on measuring the appeal of the story line, the popularity of the stars, the amount of publicity, and the reaction of preview audiences.

Gallup was convinced that 2000 people who were chosen scientifically would be a better predictor of electoral outcomes than would millions chosen in the way that the *Literary Digest* did. He conducted bi-weekly polls, which showed Roosevelt leading by increasing amounts from August through October. The result can be seen on the Gallup's organization web site, as the beginning of a continuum of similar tracking polls for US presidential elections up to the present (Figure 6.2).

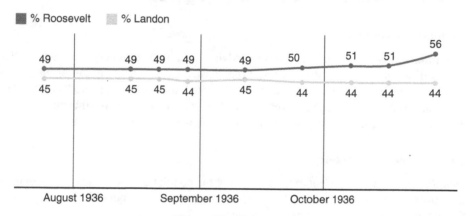

Figure 6.2   Gallup poll.

Not only did Gallup correctly predict that Roosevelt would win, he also correctly predicted the outcome of the *Literary Digest* survey. He did this via a random sample, which he had selected to replicate, as far as possible, the demographics of the *Digest* survey respondents.

The key ingredient that catapulted Gallup to fame and success was the realization that a small representative sample is more accurate than a large sample that is not representative. There are now a variety of increasingly sophisticated devices that pollsters use to ensure representative results but at the root of all of them lies random sampling.

The most fundamental form of random sampling is the *simple random sample* (SRS). What is a simple random sample?

The basic idea is that, in drawing such a sample, each element in the population will have an equal chance of being selected. For a rigorous definition, though, we need more than just the idea of "random." For example, with a population of Democrats and Republicans, we could be lazy and say that we will flip a coin and, if it lands heads, our sample will consist of all the Democrats. Each member of the population has an equal—50/50—chance of being selected, but this procedure will hardly produce a representative sample. We need more.

Definition: Simple Random Sample (SRS)

A simple random sample is produced by the equivalent of first placing the entire population, represented by slips of paper, in a box. Then, we shuffle the box and draw out the number of slips required for the sample. Statistically speaking, a sample of size n qualifies as an SRS if the sampling procedure affords each potential sample, that is, for each combination of n elements, there is an equal chance of emerging as the selected sample. The focus is on the procedure by which the sample is drawn and not on the characteristics of the resulting sample. It may therefore be more descriptive to use the term randomly drawn sample rather than random sample.

Random sampling does not guarantee a completely representative sample. In fact, the use of random sampling almost guarantees that each sample will be a little different from the population from which it is drawn. The beauty of random sampling is that we can quantify the probable extent of this difference. We will see how in a moment, but for now, let us introduce or review some key concepts of the sampling process.

Definition: Population

The **population** is the group that you are studying. It is often an amorphous concept that becomes difficult to define other than in broad terms. Consider the notion of New York voters. Does it include people who are eligible to vote but have not registered? What about out-of-state students who attend universities in New York and could vote there or at home?

Clearly, we need a working definition that we can put into practice.

Definition: Sampling frame

A *sampling frame* is a practical representation of the population—the slips of paper in the box from which we draw samples. For the New York voters, one possible sampling frame is the list of registered voters as of a given date.

Definition: Parameter

A *parameter* is a measurable characteristic of the population—for example, the mean, proportion, and so on.

Definition: Sample

A *sample* is a subset of the population. When it is randomly drawn, it is a *random sample*.

Definition: Random sampling

Technically, a random sampling process is one in which each element of a population has an equal chance of being drawn. You can think of it as a box with slips of paper that are well shuffled, and you draw slips of paper blindly.

The words "statistic" and "statistics" have several meanings, all valid in different contexts. "Statistic," in the context of sampling, is defined in the following way.

Definition: **Statistic**

A *statistic* is a measurable characteristic of a sample, and it is used to estimate a population parameter.

Random Assignment—Review

We have been talking thus far about the use of random sampling from a larger population to form representative samples. In the first course of this sequence, we spoke about the use of random *assignment* of treatments to subjects in experiments. The mechanics are similar—in each case, we can imagine a box with slips of paper and a random draw procedure.

In the case of the political survey, the goals of random sampling are to produce a sample that is reasonably representative of a larger population and to estimate the extent to which a sample estimate might be in error due to chance.

In the case of the experiment, the goal of random assignment is to ensure that any difference between the treatment groups is either due to the treatment being tested or to chance.

Later, we will look at some refinements and modifications to simple random sampling. First, though we will look at how SRSs allow us to quantify the amount of error for which we are at risk when we measure some population parameter via a sample versus taking a census of the entire population.

6.2 MARGIN OF ERROR: SAMPLING DISTRIBUTION FOR A PROPORTION

You are probably familiar with the margin of error that often accompanies survey results, such as "42% think the country is headed in the right direction with a plus-or-minus 2% margin of error."

How is the margin of error calculated? What does it mean? The answer to the first question will help you understand the answer to the second.

The margin of error quantifies the extent to which a sample might misrepresent the population it is coming from, owing simply to the luck of the draw in who gets selected in the sample.

We will start with this example.

In December 2010, a commercial polling organization sampled 200 US voters and found that only 72 voters, 36%, rated President Obama's handling of the economy positively—as good or excellent. Can we use a simulation to assess how reliable this sample is?

Before we answer this question explicitly, let's use a simulation to explore the extent to which the favorable proportion might change from resample to resample. We can put our 72 positives (1s) and 128 negatives (0s) in a box and repeatedly draw resamples of size 200, seeing how the proportion of 1s changes from draw to draw.

> ### *Try It Yourself 6.1*
>
> Using software that is capable of doing resampling simulations, execute a computer equivalent of the following simulation where 1 = "positive".
>
> 1. Put 200 slips of paper in a box. Mark 72 as 1 and 128 as 0.
> 2. Shuffle the box and draw out a number. Record the number and put the number back.
> 3. Repeat step two 199 more times and record the total number of ones.
> 4. Repeat steps 2 and 3 many times (say, 1000), recording the number of ones each time.
> 5. Produce a histogram of the results.
> 6. Without worrying about being too precise, fill in the blanks in this statement. Most of the time, the proportion who rated the handling of the economy "positively" in the sample lies between ___ and ___.

Question 6.1

What is the importance of using 200 slips of paper in step 1? Could you use, say, 36 1s and 64 2s? 18 1s and 32 0s?

Question 6.2

Which step in the above simulation is essential in modeling the size of the original sample?

You will find computer solutions to this problem using Resampling Stats, StatCrunch, and Box Sampler in the Resources and in the textbook supplements. Please read through them now.

The Uncertainty Interval

Looking at the histogram in Figure 6.3, we can see that the resample results range from about 27.0% positive to about 47.0% positive. For now, we can ignore the outliers beyond this range. We can quantify the uncertainty of the range of resample results with an interval

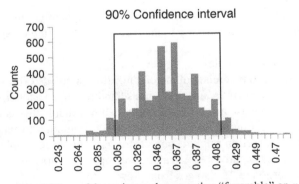

Figure 6.3   90% confidence interval; proportion "favorable" on *x*-axis.

that includes the large majority—90%, 95%, and so on—of the resample results. For example, we find this interval—called a *confidence interval*—by locating the fifth and 95th percentiles of the resampling distribution. This interval encloses 90% of the resampling results. The 90% confidence interval is approximately from 0.30 to 0.41.

To Represent the Population, How Do We Know What to Put in the Box? From our simulation mentioned earlier, we knew what was in the population, and we wanted to learn how the samples behaved. In reality, we know only the sample result—we do not know what the population holds.

What do we put in the box? We create our best guess simulated population, which is based on the observed sample—36% positive. If we wanted to have a box with all voters, that would be a box with 45 million positive slips and 80 million not positive slips.

A More Manageable Box A box with millions of slips of paper is not manageable, and it is even a bit cumbersome on the computer. We use a shortcut.

1. A smaller hat—72 slips labeled "positive" and 128 labeled "not positive."
2. Sample *with replacement.*

Sampling with replacement is equal to sampling without replacement from a huge population. The positive proportion in the box always remains pretty much the same from one draw to the next in either case, as long as the sample size remains very small, relative to the population.

The container that holds the slips of paper is variously called a *box*, an urn, or a hat. The idea is the same.

Summing Up

To produce a confidence interval,

1. We can use the observed sample as a good proxy for the population.
2. The resample size should be the same as the original sample size.
3. The fact that the sampling is done *with replacement* allows the sample to serve, in effect, as a simulated population of infinitely large size.

6.3 SAMPLING DISTRIBUTION FOR A MEAN

We just calculated a confidence interval for a proportion. Let us do it now for the mean.

When you purchase a car, the dealer typically offers to purchase your old car, which is then resold. Toyota would like to know how much those used cars sell for—this is an important piece of information in revenue projection. Let us take a simple case—establishing the average resale price of the used cars disposed of by the dealers. The Toyota regional office in Europe takes a sample of recent Corolla sales, which yields the resale values as shown in Table 6.1. The data are real sale values of used Toyotas; the scenario has been modified slightly.

Figure 6.4 is a histogram produced by StatCrunch using this procedure: My StatCrunch, > Open StatCrunch, > paste in data, Graphics > Histogram.

TABLE 6.1   Toyota Corolla used Car Prices

13,500
13,750
13,950
14,950
13,750
12,950
16,900
18,600
21,500
12,950
20,950
19,950
19,600
21,500
22,500
22,000
22,750
17,950
16,750
16,950
Mean: 17,685

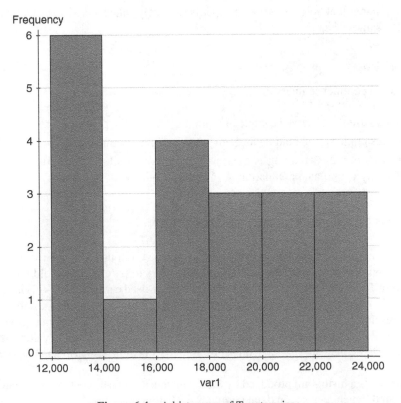

Figure 6.4   A histogram of Toyota prices.

The average sale price in this sample of 20 cars is €17,685. How much might this average be in error? In other words, if sales records could be located for all recent sales of used Toyota Corollas, how much might this estimate be of?

If we had easy access to the records of all recent sales of used Corollas, we could just compare our sample to the entire data set. But we do not have easy access—that is why we took a sample.

One way to answer this question is to actually go out and sample additional sales transactions. Take another sample of 20, another sample of 20, and so on and see how much they differ from one another. But this will cost much more time, effort, and money than taking only a single sample of 20.

Can we take additional simulated samples instead of real samples like how we did with the political poll? The trick, as always, is to determine what goes in the box. What population do we choose to sample?

Simulating the Behavior of Samples from a Hypothetical Population

Let us create a hypothetical population from our sample, remembering that we want to model a population that likely gave rise to it.

Just as above, we can replicate each item in our sample a thousand times. Unfortunately, we do not really know how big the total population of transactions is. Finding that out would take more thought and effort. Fortunately, as we saw earlier, it does not really matter as long as the sample is small relative to the population. We will end up with a population of 20,000 values, and this hypothetical population embodies everything we know about the information in the sample. We can then take resamples from this hypothetical population and see how these samples behave.

6.4   A SHORTCUT—THE BOOTSTRAP

As we have seen earlier, we do not really need a huge box. Instead, we can achieve the same effect by sampling with replacement—putting each value back into the box after we have drawn it, thereby yielding an infinite supply of each sample element.

Let us clarify some terms first.

Definition: Observation

An observation is a data value for a single case. It could be a single value or multiple values, for example, blood pressure and heart rate, for the same subject.

Definition: Sample

A sample is a collection of actual observations from a population.

Definition: Resample

A resample is a new simulated sample, that is, a collection of observations drawn from the original sample or generated randomly by a formula based on the original sample.

Definition: **Sampling with replacement**

When we sample with replacement, each item is replaced after it is drawn from a box, hat, etc.

Definition: **Sampling without replacement**

In sampling without replacement, once an item is drawn, it is not eligible to be drawn again. Sampling without replacement is also called *shuffling*.

   In the specific case described earlier, we have an original sample from a population, and instead of replicating the sample many times to create a hypothetical large population, we take resamples with replacement.

Definition: **A Single simulation trial**

A *single simulation trial* is the taking of a resample and performing further calculations with it. Typically, this means calculating the value of some statistic, such as the mean.

Definition: **A simulation**

A *simulation* is a repeat of multiple single simulation trials and the collection of their calculation results.

Definition: **The bootstrap**

In the bootstrap, we resample with replacement from an observed sample to observe the distribution of a sample statistic. This is a shortcut that eliminates the need to first replicate the values in the sample many times, then sample without replacement from that large hypothetical population.

Let's Recap

We have a sample of size N from an unknown population and we want to know how much an estimate based on that sample might be in error.

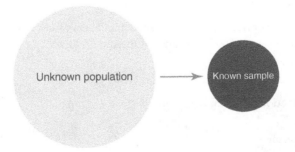

   The key question is how do samples drawn from this population behave, that is, how different are they from one another? We address this by simulating resamples of size N from a hypothetical population.

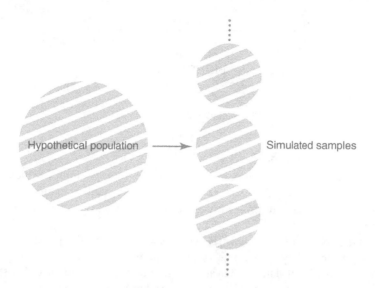

The accuracy of this procedure depends on how well the hypothetical population—the sample replicated over and over again using sampling with replacement (or the bootstrap) as the shortcut—mimics the characteristics of the unknown population.

Usually, the sample embodies all that we know about the population that it came from, and replicating it is an effective way to proceed. It is also possible to create a simulated population from the sample parameters such as a Normal population whose mean and standard deviation are estimated from the sample. See Appendix A on the parametric bootstrap in Chapter 7.

Having found an appropriate representation of the population, we can draw resamples from that population, calculating the statistic of interest and recording it each time we draw a resample. Once again, these are the steps.

1. From our observed sample, calculate a statistic from it to measure some attribute of the population that we are examining.
2. Create a hypothetical population, incorporating the best information we have. This is usually the information from the sample.
3. Draw a resample from the hypothetical population and record the statistic of interest.
4. Repeat step 3 many times.

Observe the sampling distribution of the statistic of interest to either learn how much our original estimate might be in error or how much it might differ from some benchmark value of interest.

A Bit of History—1906 at Guinness Brewery

In 1906, William S. Gossett, a chemist at the Guinness brewery firm, took a leave of absence to study with the noted statistician Karl Pearson. When he returned to Guinness, one of his concerns was the reliability of conclusions drawn from small samples. In 1908, he published an article, "The Probable Error of a Mean," in the journal *Biometrika* (Figure 6.5).

Gossett published his article under the pseudonym Student because Guinness did not want competitors to know that they were employing statisticians. The existing industry

VOLUME VI   MARCH, 1908   No. 1

BIOMETRIKA.

THE PROBABLE ERROR OF A MEAN.

BY STUDENT.

Figure 6.5   William S. Gossett's 1908 article.

practice at that time was to measure processes with very large samples, which was costly and time consuming. Being able to work effectively with smaller samples would reduce cost and speed innovation, but this required knowing something about how reliable those small samples were.

Gossett started with a simulation. He worked with data on the physical attributes of criminals. Scientific society at that time was very interested in profiling and explaining the criminal "type," so a lot of data were available from this captive audience. Gossett wanted to avoid data that would reveal that he worked for a brewery (Figure 6.6).

SECTION VI.   *Practical Test of the foregoing Equations,*

Before I had succeeded in solving my problem analytically, I has endeavoured to do so empirically. The material used was a correlation table containing the height and left middle finger measurement of 3000 criminals, from a paper by W. R. Macdonell (Biometrika, Vol. I. p. 219). The measurements were written out on 3000 pieces of cardboard, which were then very thoroughly shuffled and drawn at random. As each card was drawn its numbers were written down in a book which thus contains the measurements of 3000 criminals in a random order. Finally each consecutive set of 4 was taken as a sample—750 in all—and the mean, standard deviation, and correlation of each sample determined. The

Figure 6.6   Gossett's simulation.

Gossett then took the recorded values for his samples, plotted them, and fitted curves to them. Here is his plot of the sample standard deviations along with the curve that he fitted to them (Figure 6.7).

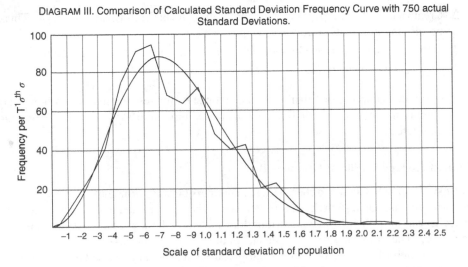

DIAGRAM III. Comparison of Calculated Standard Deviation Frequency Curve with 750 actual Standard Deviations.

Scale of standard deviation of population

Figure 6.7   Gossett's plot of SDs and fitted curve.

If computing power like we have today had been available in 1908, it seems likely that computer sampling and simulation would have played a major role in statistics from the beginning. As it was, statisticians developed mathematical approximations as shortcuts to the exact sampling distributions.

Of course, computing power is available now, and computer-intensive simulations and other procedures play a major role in statistics today. However, the legacy of formula-based mathematical approximations is still with us in textbooks and software, so we present both in this text.

6.5   BEYOND SIMPLE RANDOM SAMPLING

The procedures we have looked at so far will help us quantify random variation in SRSs.

Simple random sampling is easy in theory, but it can be complex or inefficient in practice. There are some situations when variations on simple random sampling are either easier to implement or produce better results.

Stratified Sampling

You work for an internet merchant whose web sites produce many leads. Five percent of these leads eventually becomes sales. You want to perform a survey to discover what distinguishes nonpurchasing leads from those that go on to purchase. In particular, you might want to develop a statistical model that can predict, based on data concerning the prospect, whether the prospect will end up purchasing.

Available to you is a customer database that includes a record for each customer and each prospect, that is, a contact who has expressed interest but has not purchased yet. You could take a SRS from the database. However, only about 5% of customers in the database will be purchasers. As you proceed with the survey, you will spend more time and effort to collect and examine information about a lot of nonpurchasers, while gathering relatively little information about purchasers.

An alternative is to sample equal numbers of nonpurchasers and purchasers. In this way, you will get the same amount of information about each group.

Definition: Stratified sampling

In stratified sampling, the population is split into categories, or strata (singular stratum), and separate samples are drawn from each stratum.

In the illustration shown earlier, we need equal information about purchasers and nonpurchasers, so we draw equal-sized samples from each stratum. In some cases, it may be desirable to draw different-sized samples from each stratum. For example, those who commission political polls often want to know both about the opinion of the population as a whole as well as the opinions of various subgroups (e.g., Republicans who voted for Obama). In such a case, strata sample sizes are a compromise between the need to get information about the population as a whole and the need to get information about subgroups. Taking equal-sized samples of subgroups—particularly smaller groups like Republicans who voted for Obama—diminishes the efficiency of the survey for getting information about the population as a whole.

Cluster Sampling

Let us say you want to survey a sample of high school students in a large school district. There is no readily practical and affordable way to achieve the equivalent of placing all the high school students' names in a box and drawing a sample. Various obstacles include the absence of a single student database covering all high schools and the manual labor involved in tracking down each student selected for the sample.

An easy and a practical solution is to survey the students as a group in their start-of-day administrative homeroom. A certain number of homerooms could be selected from each school, and all the students in each homeroom could be given the survey. This procedure would be reasonably representative, provided that students are assigned to these administrative homeroom units in some manner that does not introduce bias, such as by birthdate or student ID number. But if students are assigned to homerooms by, for example, athletic participation or academic ability, you run the risk of over- or under-representing groups whose opinions might be systematically different.

Definition: Cluster sampling

In cluster sampling, clusters of subjects or records are selected, and the subjects or records within those clusters are surveyed or measured. The main rationale for selecting clusters, rather than individuals, is practicality and efficiency. Care must be taken so that the characteristics that define clusters are not characteristics that will introduce bias into the results.

Systematic Sampling

Another alternative to pure random sampling is systematic sampling—the selection of every *n*th record. For example, you might sample every 100th transaction or every 10th customer. However, with systematic sampling, you lose the assurance you have with simple random sampling that the sample will be representative of the population—allowing for sampling error. So you must take care to avoid possible sources of bias. For example, if you sample daily sales and sample every seventh day, you will measure only sales on one particular day of the week. That day may not be typical, so your measurements could be biased.

Multistage Sampling

Professional polling and sampling organizations use a mix of methods to minimize cost, sampling error, and bias. These organizations have the benefit of having a lengthy experience with different methods. Therefore, they can assess from their prior work whether a departure from simple random sampling, in the name of efficiency, will introduce bias. An example of a multi-stage process might be to randomly select neighborhoods, as defined by the US census, and then sample every 10th household.

Convenience Sampling

You have probably encountered polls on the web in which you are invited to state your opinion on something. This is an example of convenience sampling. In convenience sampling, there is no effort to define a population or sampling frame, and there is no attempt to assure that the sample is representative of a larger population. It is simply taking a sample on the basis of convenience. This sampling method is easy and cheap, but it does not produce consistently useful or reliable information.

Self-Selection

An added problem with the convenience sampling method is that the sample is self-selecting. The organization collecting the data is not selecting the sample; the respondents themselves determine whether they participate. In an opinion survey, this almost guarantees biased results—those with stronger opinions are more likely to participate.

The difficulty with self-selection can be seen in web-based product reviews. Consider these two reviews of a printer from Amazon, January 31, 2011.

By Lady ☑ - See all my reviews
This review is from: **HP Deskjet 3050 All-in-One Printer (CH376A#B1H) (Electronics)**
I bought this printer to use specifically for wireless purposes with my laptop. Since day one it has not worked properly. Sometimes it prints right away and other times it takes 45 min to print one page! Its very frustrating. Even when I sit right next to my router and printer I have issues. Funny thing, today I printed 30

By John W. Bradshaw "jwbradshaw" ☑ (Allen, TX) - See all my reviews
REAL NAME
Amazon Verified Purchase (What's this?)
This review is from: **HP Deskjet 3050 All-in-One Printer (CH376A#B1H) (Electronics)**
This dynamic printer is small, easy to use, fast and produces a clean, high quality classy print, whether it is text or enlarged, detailed photograph. You can use direct connection (USB) or wireless connection. The instructions and disc that come with the printer make it a snap to set up and begin printer. Even the wireless was hands-free and automatic. For its size and versatility, this printer is unmatched in price and speec of delivert. Very highly recommended!

Because it appears that anyone who wants to can post a review, the survey method makes it difficult to draw conclusions. Are disgruntled customers more likely to post reviews and skew the results? Does the vendor have an incentive to write spurious positive reviews? What would motivate an ordinary user who has had no particular issues to write a review? Are ordinary users who write reviews different from those who do not write reviews?

Yelp

Yelp, the community recommender service, has a significant stake in the validity of its community reviews (it is their main product). The company has procedures in place to prevent "promotional" reviews from seeing the light of day. One such procedure is to down-weight reviews coming from those who are not long-time members of the Yelp community with other reviews to their credit. Among some merchants who are the subject of negative reviews, though, a different perspective has taken root, they consider that Yelp is gaming the system, showing more positive reviews for merchants who advertise with Yelp.

Nonresponse Bias

Nonresponse bias is a problem that can occur no matter what the sampling method is. It is the question of whether people who respond to a survey are different from those who do not. As nonresponders by definition do not respond and do not show up in surveys, we cannot know for sure if they are different from responders. However, it is hard to rule out this possibility.

Question 6.3

How will a survey response rate affect the problem caused by nonresponse bias?

Estimating sampling error in these variations cannot easily be done via formulas; resampling approaches that are beyond the scope of this book must often be used.

6.6   ABSOLUTE VERSUS RELATIVE SAMPLE SIZE

Let us return to the story about the Gallup Poll versus the *Literary Digest*. Gallup believed that selecting a representative sample was more important than selecting a large sample. The *Literary Digest* touted the millions of people in its survey.

Let us consider a related issue. Consider these two questions.

1. What size sample do you need to obtain accurate voter preference estimates for the United States, which has a population of 300 million?
2. What size sample do you need to obtain accurate voter preference estimates for Albany, NY, with a population of 300 thousand?

When asked these questions, many people think that a larger sample is required for the United States than for Albany. Is this correct?

Try It Yourself 6.2

Suppose a survey of 1000 people about Obama's handling of the economy is conducted for Albany, as well as for the United States as a whole, spell out the simulation steps, as mentioned earlier, that would be required to assess its accuracy. For convenience, we can stick with 36% positive. To achieve the same degree of accuracy for the United States, is a larger sample required?

6.7   EXERCISES

Use the PulseNew.xls data for the first two questions:

1. Suppose you want to estimate the proportion of all students who smoke, would that be an experiment or a survey? For estimating this proportion, is the PulseNew.xls data a good (survey) (experiment) or a bad one? Why?

2. If the goal of this study were to estimate the effect of running in place on pulse rate, would it be an experiment or a survey? Would it be a good one or a bad one? Why?

3. You work for a survey organization that has been asked to conduct a survey to determine what proportion of football players suffer from health conditions related to head

trauma. The client seeking the survey has not been more specific than that. You are developing a plan that can serve as a starting point to discuss exactly what the client wants. Identify

(a) an appropriate population

(b) the sampling frame

(c) any other practical issues

4. An online retailer is interested in learning its median transaction size for the previous year. Describe a possible sampling procedure that would avoid having to look at all transactions. (Do not use statistical terms; instead write down a series of steps that, for example, a summer intern could follow.)

5. Consider the following populations:
A = { 1 2 3 3 }
B = {one million 1's, one million 2's, two million 3's}

(a) If you randomly draw one value without replacement from population A, what is the probability of drawing a "3"?

(b) If you randomly draw two values without replacement from population A and draw a "3" on your first draw, what is the probability of drawing a "3" on the second draw?

(c) If you randomly draw one value without replacement from population B, what is the probability of drawing a "3"?

(d) If you randomly draw two values without replacement from population B and draw a "3" on your first draw, what is the probability of drawing a "3" on the second draw?

(e) If you randomly draw one value with replacement from population A, what is the probability of drawing a "3"?

(f) If you randomly draw two values with replacement from population A and draw a "3" on your first draw, what is the probability of drawing a "3" on the second draw?

(g) Consider procedures b, d, and f. Which of them are almost exactly equivalent to one another?

6. The "churn" rate in a subscription business is that rate at which subscribers leave in a given time period. In the US wireless industry, the churn rate is on the order of 10% annually—out of 100 customers at the beginning of a 12-month period, only 90 will be left at the end of the 12-month period.

(a) A 10% churn rate can be represented by a single box with cards marked 0 and 1. Different sized boxes (i.e., different numbers of cards) can also do the job—give several examples.

(b) What resampling process makes these different boxes functionally equivalent?
   – sampling with replacement
   – sampling without replacement

(c) Describe the resampling steps that you would take to assess the sampling variability of samples of size 250.

(d) Describe the resampling steps that you would take to assess the sampling variability of samples of size 50.

Note: The answers above can be in the form describing the steps in a box model, such as

– put slips of paper in a box
– take a sample
– etc. (only more detailed).

7. A politician contemplating running for local office wants to know how widely his name is recognized and undertakes a sample survey of 200 potential voters. Sixty-five of them had heard of him, the others had not.

   (a) Describe the steps you would take in a simulation to determine the reliability of this sample result.

   (b) Use Resampling Stats, Box Sampler, or StatCrunch to carry out the simulation.

8. A politician contemplating a run for state office suggests to his campaign consultant that, instead of using expensive surveys to measure public opinion, they should simply harvest Twitter feeds and use sentiment analysis. Comment.

Answers to Try It Yourself

6.2 If you think it through, you will realize that a sample of 1000 is tiny relative to the population in both Albany and the United States as a whole. The simulation procedure used above is the same in both cases:

1. A hat with 36 "positive" and 64 "not positive" slips of paper.

2. Draw a slip of paper, record the result, and put the slip of paper back.

3. Repeat step 3 1000 times (a single resample), record the proportion "positive" in that resample.

4. Repeat steps 2 and 3 many times (say 5000).

5. Display the results in a histogram so you can see the distribution of proportion "positive."

The accuracy of this procedure (the sampling variability) is the same, whether it is applied to Albany or to the United States as a whole.

Conclusion: You do not need a larger sample size for the United States as a whole than for Albany. To achieve a given degree of accuracy in sampling opinion, the same size sample should be used for both Albany and the United States as a whole. A larger sample will give more reliable results, but this will be true in equal measure for both Albany and the United States as a whole.

Answers to Questions

6.1 We started with 72 1s and 128 0s because that is what we got in our original sample. You could achieve the same effect with, say, 36 1s and 64 0s or 18 1s and 32 0s. As you are drawing with replacement, the probability of drawing a "1" will remain the same whatever the size of the box. As long as it has 36% 1s, it will properly model the problem (when dealing with proportions and percentages, using 100 slips of paper can be helpful a 100 matches nicely to percentages.)

6.2 The original sample size was 200. Step 3 is the essential step in modeling the sample size—it determines the size of the resample being drawn from the box.

Why not Step 1, in which we created the box with 200 slips of paper?

Because it is not the size of the box that models sample size behavior but the size of the resample that you draw from the box. Recall that we actually could have used a box with 100 slips of paper or some other number as long as 36% of them were 1s. As long as we resample with replacement, it does not matter—the probability of drawing a 1 remains unchanged.

But the size of the resample does matter—the proportion of 1s in a small resample will vary a lot more than in a large resample. If we want to assess variability in our original sample, the resample size must equal the original sample size.

6.3 The lower the response rate, the greater the exposure to nonresponse bias. Suppose that 10% of the purchasers of a particular product are unhappy and have a much greater propensity to respond to a customer satisfaction survey. If 75% of the purchasers respond, the disgruntled 10% will not have a great impact on the results. If, however, only 20% respond, the disgruntled 10% will dominate.

The best way to deal with nonresponse bias is to maximize the response rate, mainly through two avenues:

- Making the survey quick and easy
- Following up with those who do not respond

Money spent on follow-up to ensure a high response rate is often better spent than money spent expanding the sample size.

There are other methods that statisticians have developed to adjust for nonresponse bias, but they are beyond the scope of this course.

7

CONFIDENCE INTERVALS

In this chapter, we take the simulation procedures that we have been working on and put them in a more formal (traditional) statistics framework. After completing this chapter, you should be able to:

- distinguish between the appropriate uses of point estimates and interval estimates,
- calculate confidence intervals (via resampling or formulas),
- explain the relationship between the Central Limit Theorem and the applicability of Normal approximations for confidence intervals,
- calculate standard error and explain the difference between it and standard deviation,
- calculate the confidence interval for a mean or proportion,
- calculate the confidence interval for a difference in means or proportions.

This material, particularly the vocabulary and definitions, is most relevant for the *research* community. *Data scientists*, however, will encounter confidence intervals in their work and will benefit from a solid understanding, via resampling, of how they work.

7.1 POINT ESTIMATES

The procedures we have discussed all involve establishing the possible error that occurs when we measure some parameter in a population by taking a sample from that population.

Introductory Statistics and Analytics: A Resampling Perspective, First Edition. Peter C. Bruce.
© 2015 John Wiley & Sons, Inc. Published 2015 by John Wiley & Sons, Inc.

The technical term for this procedure of establishing the possible error is a confidence interval. It is one way to measure the accuracy of a measurement. The statistic or the measurement itself is often called a *point estimate*.

Definition: Point estimate

A point estimate is a statistic, such as a mean, median, and percentile from a sample. The term "point" is to emphasize the fact that it is a single value and not a range of values.

7.2   INTERVAL ESTIMATES (CONFIDENCE INTERVALS)

We have seen that individual sample results can vary, and a point estimate has an uncertainty attached to it. Often, it will be misleading to report simply the point estimate. Instead, we will report an interval, to account for the uncertainty.

Earlier, we discussed a political survey that asked 200 people their opinion of President Obama and 72 were favorable. If we ask another 200, we would be a bit surprised if, again, exactly 72 were favorable. There is some uncertainty that arises from the luck of the draw in sample selection. We quantify that uncertainty with a confidence interval, which is used to bracket a point estimate from a sample.

A confidence interval is similar to the margin of error that we saw earlier. You might hear the following on the news: In a poll, the president's approval rating was 57% plus or minus two percentage points. The point estimate is 57%. The margin of error is plus or minus 2 percentage points. The confidence interval is 55–59%.

In addition to the interval endpoints, a confidence interval is always characterized by a percentage value such as 90%. This is the percent of time that the resampling results we have been discussing fall within that interval.

Definition: A 90% confidence interval for the mean (or other statistic)

A 90% confidence interval is a range that encloses the central 90% of the resampled means (or other statistic), using the simulation procedures described in the following sections. A similar interval can also be produced by the formula-based alternatives also described below.

A more technical definition of a 90% confidence interval is that it is an interval that would enclose the true statistic 90% of the time when constructed repeatedly in the same manner with the same population.

Confidence Interval vs Margin of Error

In surveys, the term "margin of error" is frequently used. A margin of error is simply a plus or minus quantity attached to a point estimate, whereas a confidence interval is the actual endpoints of the interval. Typically, the confidence interval is constructed and then a margin of error is calculated from it. This works if the confidence interval is symmetric about the point estimate (not always the case). Suppose that a survey estimate is "48% yes" for some proposition.

If the confidence interval is 44–52%, then the margin of error is 48% ± 4%.

7.3   CONFIDENCE INTERVAL FOR A MEAN

We have actually already calculated a confidence interval for a mean, in effect. Let us review the Toyota Corolla case described earlier, where the average price in the sample—the sample mean—is €17,685.

Resampling Procedure (Bootstrap):

1. Write all 20 sample values on slips of paper and place them in a box.
2. Draw a slip from the box, record its value, and replace the slip.
3. Repeat step two 19 more times and record the mean of the 20 values as shown in Figure 7.1.
4. Repeat steps two and three 1000 more times.
5. Arrange the 1000 resampled means in descending order and identify the fifth percentile and the 95th percentile—the values that enclose 90% of the resampled means. These are the endpoints of a 90% confidence interval, as shown in Figure 7.2. Figure 7.3 is a histogram of the 1000 resampled means.

You can view three video tutorials that illustrate the Toyota resampling or bootstrap procedure on a computer by clicking these links.

For a Resampling Stats video tutorial, see the file Toyota_rsxl.html.

For a Box Sampler video tutorial, see the file Toyota_box.html.

For a StatCrunch video tutorial, see the file Toyota_statcrunch.html.

Results: In one set of 1000 simulation results, the interval was from €16,415 to 18,860. So, we would report these sample results as follows:

The estimated price of a used Toyota Corolla is €17,685 with a 90% confidence interval ranging from €16,415 to 18,860.

7.4   FORMULA-BASED COUNTERPARTS TO THE BOOTSTRAP

Computational power and fast statistical software were not widely available until the 1980s, so the previous simulations were not feasible until then. Statisticians instead developed approximations that allowed analysts to calculate confidence intervals from formulas. These formulas are still used in software and books, and so we present them here. The binomial and Normal distributions play important roles in these calculations. We introduce the procedures using the Normal distribution.

Formula or resampling?

 In the *research* community, those planning to study more advanced statistics will encounter these formula counterparts to the resampling operations but generally need not do calculations by hand, but by using software instead.

Data scientists are not generally concerned with formal statistical inference based on formulas but should pay attention to resampling procedures to get familiar with random chance models and to gain practice with coding statistical algorithms "on demand."

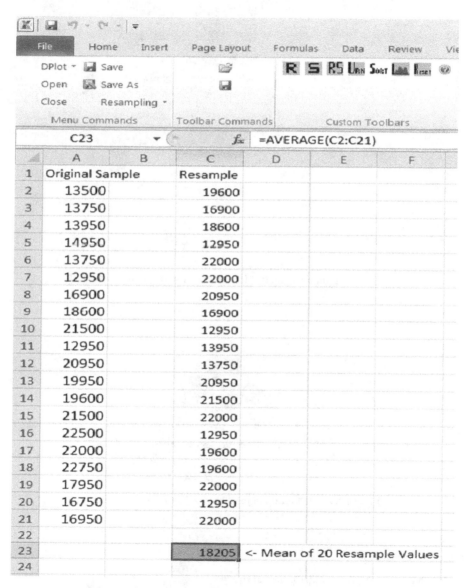

Figure 7.1   Mean of 20 resampled or bootstrapped values.

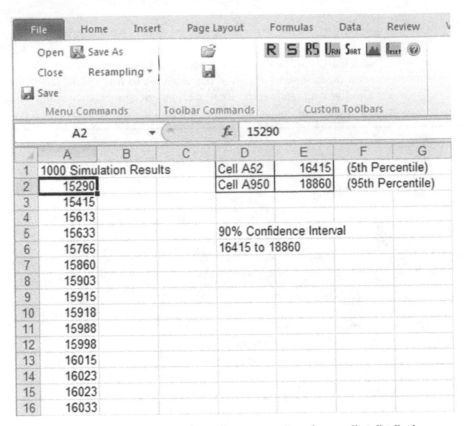

Figure 7.2   90% confidence interval from percentiles of resampling distribution.

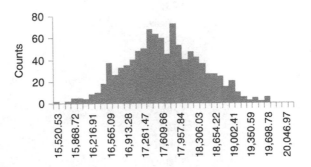

Figure 7.3   Histogram of used Toyota Corolla resale values.

Normal Distribution

We introduced the Normal distribution earlier.

In that discussion, we explained the distribution of a population or sample data. Naturally occurring Normally distributed data are not as common as you might initially think. Here are some examples.

- *Heights*: On average, men are taller than women, children are shorter than adults, and different ethnic groups can vary in their heights. Only when you control for such factors and restrict the definition of the group, for example, black American adult females, does the distribution turn normal.
- *Incomes*: The distribution of incomes is long-tailed or skewed to the right to account for the individuals who have incomes many times greater than the average.
- *Tech support call duration*: Like incomes, this distribution is right-skewed, reflecting the fact that there are a few extremely difficult and lengthy calls.

Central Limit Theorem

Despite cases like those given earlier where population and sample data are not normally distributed, the Normal distribution may still apply for the *means* of samples drawn from those populations. Depending on the size of the sample and the degree of non-normality in the parent population, sample means are often Normally distributed. This phenomenon is termed the *Central Limit Theorem*.

Definition: **Central limit theorem**

The Central Limit Theorem says that the means drawn from multiple samples will be Normally distributed, even if the source population is not Normally distributed, provided that the sample size is large enough and the departure from Normality is not too great.

Many books state that the "large enough" range is a sample size of 20–30, but they leave unanswered the question of how non-Normal a population must be for the Central Limit Theorem to not apply.

The Central Limit Theorem allows Normal-approximation formulas to be used in calculating sampling distributions for inference, that is, confidence intervals and hypothesis tests. With the advent of computer-intensive resampling methods, the Central Limit Theorem is not as important as it used to be because resampling methods for determining sampling distributions are not sensitive to departures from normality.

Try It Yourself 7.1

Check out this demo of the Central Limit Theorem http://onlinestatbook.com/stat_sim/sampling_dist/index.html.
Click on the "Begin" button and you will see four graph areas.

1. The top area is a frequency histogram of the population. You can choose the population shape with a dropdown. If you choose "custom," you can draw populations of different shapes.

2. The next graph is for a histogram of a simulated sample drawn from that population. You can choose how many simulated samples to draw—5; 1000 or 10,000.
3. The third graph shows the frequency distribution of sample means and other statistics. The sample size is selected here—for example, $N = 20$.

Exercise

Choose "custom" for the population, use your mouse to draw differently shaped populations and choose "sample size $N = 25$." Then click on 1000 in the "Sample" box to run the simulation. Comment on the shape of the distribution of means compared to the shape of the population.

Formula: Confidence Intervals for a Mean—z-Interval

To calculate a z-interval, we revisit the standard Normal distribution that we saw earlier. Procedurally, these are the steps.

1. Find the values in a standard Normal distribution that correspond to the fifth and 95th percentiles. These values are -1.6449 and $+1.6449$.
2. Multiply by the sample standard deviation divided by the square root of the sample size. Then, add the sample mean.

Hint: If you Google Normal distribution probabilities, you will find web-based calculators for the above. With many such calculators, you can enter the sample standard deviation and mean and then obtain the interval directly.

The shaded area in Figure 7.4 represents the 90% z-interval for a standard Normal distribution, that is, a normal distribution with mean $= 0$ and standard deviation $= 1$. We map this distribution to our sample by scaling it (using the sample standard deviation) and shifting it (adding the sample mean).

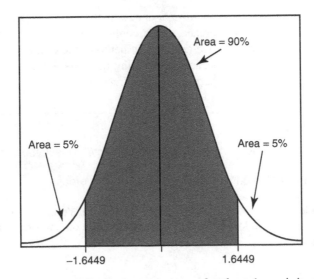

Figure 7.4   Normal distribution with mean $= 0$, stdev $= 1$; x-axis is z-score.

For a Mean: T-Interval

The Normal distribution is actually not the best approximation to the actual distribution of the sample mean when samples have fewer than 30 values. Gosset's 1908 article in *Biometrika* (see earlier) contained a superior approximation that is now known as *Student's t*. The *t*-distribution is actually a family of distributions, whose shapes differ depending on the size of the sample. The *t*-distribution has the same general shape as the Normal and is almost indistinguishable for samples of size 30 or more. As sample sizes diminish, the *t*-distribution takes account of the greater variability with small samples and becomes lower and longer-tailed than the Normal. As there are multiple *t*-distributions to be used when there are different sample sizes, a parameter called *degrees of freedom* must be specified when using the *t*-distribution. The number of degrees of freedom for the *t*-distribution = the sample size minus one.

$$\text{The } 100*(1-\alpha)\% \text{ } t\text{-interval for the mean } =$$

$$\left(\bar{x} - t_{n-1,\,\alpha/2} * \frac{s}{\sqrt{n}} \, , \bar{x} + t_{n-1,\,\alpha/2} * \frac{s}{\sqrt{n}} \right)$$

where n is the sample size, $\bar{x}$ is the sample mean, s is the sample standard deviation, $n-1$ are the degrees of freedom, and $-t_{n-1,\,\alpha/2}$ and $t_{n-1,\,\alpha/2}$ are the t values corresponding to the $\alpha/2$ percentile and the $1-\alpha/2$ percentile, respectively.

Example—Manual Calculations

The *t*-interval calculations for the Toyota example are shown in the following section.

The degrees of freedom for the one-sample t test are $n-1=19$.

If $\alpha = 0.1$, then the fifth and the 95th percentiles for t_{19} are -1.7291 and 1.7291, respectively. These values, which used to be found in tables, are now readily found in web calculators.

Using the same measures of $\bar{x} = 17{,}685$ and $s = 3507.252$ from the previous example, the 90% *t*-interval for the mean is

$$= \left(\bar{x} - t_{19,.05} * \frac{s}{\sqrt{n}} \, , \bar{x} + t_{19,.05} * \frac{s}{\sqrt{n}} \right)$$

$$= \left(17{,}685 - 1.7291 * \frac{3507.252}{\sqrt{20}} \, , 17{,}685 + 1.7291 * \frac{3507.252}{\sqrt{20}} \right)$$

$$= (17{,}685 - 1356, \ 17{,}685 + 1356)$$

$$= (16{,}329, \ 19{,}041).$$

Example—Software

In most cases, you will produce an interval using software as part of the data analysis. The following 90% confidence interval was produced with StatCrunch (My StatCrunch, > Open StatCrunch, > paste in data, Stats > *T*-statistics, One-sample, with data):

90% Confidence Interval Results

μ: mean of variable

Variable	Sample Mean	Std. Err.	DF	L. Limit	U. Limit
var1	17,685	784.24536	19	16,328.936	19,041.064

The result is similar to but not exactly the same as the bootstrap interval.

7.5 STANDARD ERROR

Another statistic that is used to characterize the reliability of a sample result is the *standard error* (abbreviated as SE).

Definition: **Standard error**

The standard error of a sample statistic is the *standard deviation* of that sample statistic. It is often termed the "*standard error of the estimate*," where "sample estimate" means the same thing as "sample statistic."

Note that the term standard deviation as used earlier refers to variation in the sample *statistic* and not to variation in the sample *data*.

Let us illustrate with the results of the Toyota bootstrap, where the sample statistic of interest was the mean. We will use the 1000 resampled means, as earlier, but instead of finding percentile intervals, we will calculate the standard deviation of the means.

1	Resampled means (1000 simulation trials)				
2					
3	17765		774.7063	= standard deviation of resampled means	
4	17997.5				
5	18250				
6	17407.5				
7	18190				
8	17975				
9	17882.5				

In calculating the standard deviation, the data are the 1000 means, and $n = 1000$. If we had done 10,000 trials, then n would equal 10,000. We divide by $n - 1$ in the calculation, although dividing by n would make little difference.

Standard Error Via Formula

The standard error can also be estimated using the standard deviation of the *sample data* itself. Here is the formula for the standard error of the mean, where s refers to the standard deviation of the sample values and n to the sample size.

$$SE_{\bar{x}} = \frac{s}{\sqrt{n}}$$

7.6   CONFIDENCE INTERVALS FOR A SINGLE PROPORTION

The procedure we followed for the margin of error in a sample survey is the confidence interval for a proportion. Let us look at another example.

An online merchant, in an effort to reduce returns, starts a pilot program in which it provides additional explanatory information and pictures about several products. Out of the next 20 purchases for those products, four are returned.

Question 7.1

What is the point estimate for the return rate under the new program?

Question 7.2

Determine a confidence interval around this point estimate.

Resampling Steps

We want to determine how samples of size 20 might differ from one another. We do not know what the returned proportion is in the population, so we use the sample proportion, 0.20, as our best guess. Then, we draw resamples from a population box that contains 20% returned. The products being sold by the company, worldwide and over time, constitute a very large population, much larger than 20, so we resample with replacement. This reflects the fact that the probability that a product will be returned is constant from one sale to the next.

(1) We can represent the 20% returned by a box with four ones, that is, returns, and 16 zeros, that is, not returned.

(2) Draw a number from the box and record whether it is a zero or a one. Replace the number.

(3) Repeat step two 19 more times, for a resample of 20 from the box. Record the number of returns—ones.

(4) Repeat steps two and three 1000 times, recording the number of ones in the resample each time.

(5) Order the results and find the fifth and 95th percentiles. This is a 90% confidence interval for the number of products returned. Divide by 20 and multiply by 100 to get percent returned.

Specific resampling procedures for this example using StatCrunch, Box Sampler, and Resampling Stats can be found in the textbook supplements.

Binomial Distribution

We just used resampling to determine how the number of ones in a sample might differ from sample to sample. For relatively small samples, this information can also be calculated using the binomial formula and distribution. For larger sample sizes, the Normal distribution is a good approximation.

Question 7.3

What assumption are we making here about independence?

 Analysts often assume independence when it is not warranted because without that assumption, analysis becomes difficult or impossible. But no analysis may be better than bad analysis!

The financial collapse of 2008 was prompted, in part, by poorly designed financial products that were based on bundles of subprime mortgages. The default risk of these bundles was confidently predicted to be at a comfortably low level, based partly on the assumption that the failure of mortgage A did not affect the probability that mortgage B would fail. In fact, this was not strictly true—systemic problems, including fraud and gross negligence, affected most subprime mortgages, and their outcomes were quite correlated.

The *binomial formula* tells you the probability of getting exactly x successes in n trials, when the probability of success on each trial is p.

Example: A baseball batter has a 0.3 probability of getting a hit in each at bat. If we observe five at-bats, the probability that he will get exactly three hits is calculated by the binomial formula to be 0.132.

It is rare that you would need to calculate this by hand; normally, it would be done by a statistical software or a web calculator. You will do that in a moment. For completeness, the formula is presented in Appendix C: Binomial Formula Procedure.

Try It Yourself 7.2

Use your chosen software to find the above probability.

There are example procedures using Excel, Resampling Stats, StatCrunch, and Box Sampler in the textbook supplements.

Often you want to know the probability of getting a range of outcomes—say, "two or fewer hits." The binomial distribution shows the probability of all possible outcomes for a given probability and the number of trials.

Example: Here is the distribution of success probabilities when $n = 5$ and $p = 0.3$ (Figure 7.5):

Probability of different successes

Successes	Probability
0	0.168
1	0.36
2	0.309
3	0.132
4	0.028
5	0.002

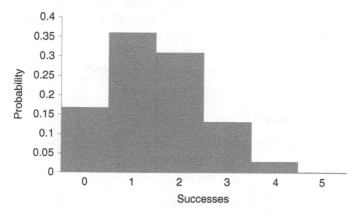

Figure 7.5   Probability of successes (hits).

Question 7.4

Answer the following questions based on the information in this table:

1. *What is the probability of getting four successes in five trials?*
2. *What is the probability of getting two or fewer successes?*
3. *How would you create the table shown earlier using resampling?*

Try It Yourself 7.3

Use your chosen software to find the probability of 3 or more successes in five trials, when the probability of success on each trial is 0.3.

Multiplication Rule (An Aside)

Recall that the multiplication rule states that the probability of independent events A AND B happening is P(A) times P(B). This also holds for multiple events A, B, C, and so on, so the multiplication rule can be employed for the extreme cases of no successes and five successes. "No successes" is really "no success on first trial" AND "no success on second trial" AND "no success on third trial," and so on.

Question 7.5

Use the multiplication rule to calculate

(1) The probability of no successes.
(2) The probability of five successes.

Normal Approximation

For large sample sizes, binomial calculations may not be available in your particular software. In such cases, ranges of outcomes are of greater interest than the probability of a specific outcome. For example, a survey where $n = 1500$ might return a result of 820 "favorable opinion of the president," and you want to establish a margin of error around 820. In such cases, the Normal distribution and the area under the Normal curve are used instead to provide a good approximation. See Appendix C: Binomial Formula Procedure.

These are Alternate Approaches

The resampling approach, the binomial formula and the Normal-approximation formula are alternate methods. You can use any of them.

7.7   CONFIDENCE INTERVAL FOR A DIFFERENCE IN MEANS

A major retailer considers switching vendors for its ultrasonic humidifiers. Vendor A is the current supplier and vendor B is offering a lower wholesale price and better commercial terms overall. Before deciding to switch, however, the retailer decides to test the efficiency of the two products—the rate at which they humidify the air. The retailer decides to compare a sample of humidifiers from each vendor, with the results shown in Table 7.1.

TABLE 7.1   Humidifier Moisture Output (oz./h)

Vendor A	Vendor B
14.2	12.4
15.1	13.6
13.9	14.6
12.8	12.8
13.7	11.9
14.0	13.1
13.5	12.8
14.3	13.2
14.9	14.7
13.1	14.5
13.4	
13.2	
Mean: 13.84	Mean: 13.36

Vendor B's average output is about 1/2 oz. less than Vendor A's.

This example is based on a similar one in T. Ryan's *Modern Engineering Statistics* (Wiley Interscience, Hoboken, 2007), which was, in turn, drawn from a study by Nelson in the *Journal of Quality Technology* (v. 21, No. 4, pp. 232–241, 1989).

Question 7.6

What decision should the retailer make? Stick with A? Switch to B?

This result—a difference of about 1/2 ounce—is a point estimate based on the two samples. How much might it be in error, based simply on the luck of the draw in selecting humidifiers for the sample? If we or the decision-maker base a decision solely upon the point estimate, it is not a complete picture. We need to couple the point estimate with a confidence interval that brackets it.

The confidence interval procedure asks "How would this result differ if we drew many additional samples?" It is not feasible to draw lots of new samples, but we can draw resamples and observe their behavior.

Resampling Procedure—Bootstrap Percentile Interval

In constructing a confidence interval for this problem, that is, to have the resampling world replicate the real world, we will want to resample from *two* boxes—one box for vendor A and one box for vendor B. (In the next chapter, we will see that if we were doing a hypothesis test, we would want just one box representing the imaginary world of a single population whose behavior we would want to test.) We will calculate a 90% confidence interval.

1. Box A has 12 slips of paper with the 12 values for Vendor A
2. Box B has 10 slips of paper with the 10 values for Vendor B
3. Draw a sample of 12 with replacement from Box A and record the mean.
4. Draw a sample of 10 with replacement from Box B and record the mean.
5. Record the difference—Mean A minus Mean B
6. Repeat steps three through five 1000 times.
7. Review the distribution of the 1000 resampled means by creating a histogram and find the 5th and 95th percentiles. These are the bounds of a 90% confidence interval. See Figure 7.6 for the histogram our example produced. Specific software procedures for this example using Resampling Stats, StatCrunch, and Box Sampler can be found in the textbook supplements.

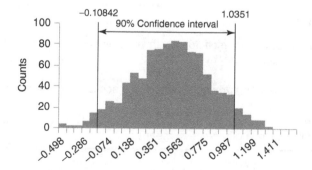

Figure 7.6   Histogram with 90% confidence interval—mean of A minus mean of B.

Question 7.7

Could the resampling procedure do the subtraction the opposite way? Mean B minus Mean A?

Question 7.8

Why draw from two boxes instead of one?

Formula—Confidence Interval for a Difference in Means

The conventional procedure for drawing a confidence interval around the difference in means uses the same t-distribution developed by Gossett that we saw.

We will denote the sample sizes of the two samples as n_1 and n_2, the sample means as $\overline{X}_1$ and $\overline{X}_2$, and the sample variances as s_1^2 and s_2^2. We further assume that the two samples are independent—not paired. The lower and upper bounds of a $1 - \alpha$ confidence interval are calculated below.

Lower bound:

$$\overline{X}_1 - \overline{X}_2 - t_{v, \alpha/2} \, ^* \sqrt{\frac{s_1^2}{n_1} + \frac{s_2^2}{n_2}}$$

Upper bound:

$$\overline{X}_1 - \overline{X}_2 + t_{v, \alpha/2} \, ^* \sqrt{\frac{s_1^2}{n_1} + \frac{s_2^2}{n_2}}$$

The degrees of freedom of Student's t-distribution are represented by v, the Greek letter nu. Degrees of freedom can be found with this equation.

$$v = \min(n_1, n_2) - 1$$

You may also see references to calculations using a "pooled variance." Such calculations assume that the two samples, ostensibly coming from two different populations, share the same variance. The circumstances under which this would be true are sufficiently limited that the pooled variance case is not covered here.

Try It Yourself 7.4

Try working this problem yourself to obtain a confidence interval for the difference between the two vendors, either by

1. Replicating the resampling procedure (there is a section in the textbook supplements on this), or
2. Using the formula approach, via software such as StatCrunch that has an option for "confidence interval for a difference in means."

7.8 CONFIDENCE INTERVAL FOR A DIFFERENCE IN PROPORTIONS

The connection between cholesterol and heart disease was first suggested back in the 1960s. You can calculate your own risk of a heart attack in the next 10 years at this website: http://hp2010.nhlbihin.net/atpiii/calculator.asp. Cholesterol levels are a key driver in these calculations.

Kahn and Sempos (*Statistical Methods in Epidemiology*, Oxford Univ. Press, New York, 1989, p. 61) describe some early research that quantified this connection. Men with high cholesterol were found to have heart attacks at a rate that was 64% higher than men with low cholesterol (Table 7.2).

TABLE 7.2 Cholesterol and Myocardial Infarctions (MI)

Cholesterol Level	MI	No MI	Total	Proportion
High	10	125	135	0.0741
Low	21	449	470	0.0447
Total	31	574	605	0.0512

The difference in proportions is:

$$0.0741 - 0.0447 = 0.0294$$

In percent: 2.94%

The high cholesterol group's risk of heart attack was 2.9 percentage points higher than the low cholesterol group. In relative terms, this is a major difference, as it is almost 65% higher.

How much might this be in error, based on sampling variation? Find a 95% confidence interval.

Resampling Procedure

Because Box Sampler is limited to a sample size of 200, this simulation must be done using either Resampling Stats for Excel or StatCrunch.

1. In one box—the high cholesterol box—put 10 slips of paper marked 1 for heart attacks and 125 slips marked 0 for no heart attacks.
2. In a second box—the low cholesterol box—put 21 slips of paper marked 1 and 449 slips marked 0.
3. From the first sample, draw a resample of size 135 randomly and with replacement. Record the proportion of ones.
4. From the second sample, draw a resample of size 470 randomly and with replacement. Record the proportion of ones.
5. Record the [result from step three] minus the [result from step four].
6. Repeat steps three through five 1000 times.

7. Find the interval that encloses the central 95% of the results—chopping 2.5% off each end. Figure 7.7 illustrates this interval. Specific software procedures for this example using Resampling Stats and StatCrunch can be found in the textbook supplements.

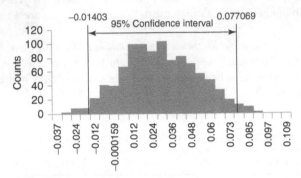

Figure 7.7   Histogram with 95% confidence interval, difference in proportion 1s, resample group of 135 minus resample group of 470.

We read the earlier-mentioned confidence interval as follows: The original study shows that men with high cholesterol suffer heart attacks at a rate that is 2.9 percentage points higher than men with low cholesterol; the 95% confidence interval runs from 7.7 percentage points higher to −1.4 percentage points lower.

Note that subsequent studies confirmed the original result, and it is now pretty well established that cholesterol and heart disease are related in ways that chance cannot explain.

7.9   RECAPPING

The vocabulary and details of statistical inference can be confusing and appear disconnected from applications, so this chapter merits a brief summary:

1. Learned three roughly equivalent ways to calculate a confidence interval for a single proportion:
   • Resample 0s and 1s from a box, recording the proportion of 1s each time.
   • Use the binomial formula.
   • Use the Normal approximation to the binomial.

   All three methods are most likely to be deployed via software; details of the binomial formula and the Normal approximation have been described in an appendix.

2. Learned two roughly equivalent ways to calculate a confidence interval for a difference between two means:
   • Resample the actual values separately from two boxes, recording the difference in resample means.
   • Use the *t*-distribution.

3. Learned how to calculate a confidence interval for a difference in proportions via resampling:
   • Resample 0s and 1s from two separate boxes, recording the difference in proportion of 1s.

The formula approach for this, the Normal approximation to the binomial, is presented in an appendix.

APPENDIX A: MORE ON THE BOOTSTRAP

Interested in learning more about the relationship between resampling and traditional formulas? Read on.

The advantage of the basic bootstrap technique described earlier is its simplicity. The hypothetical population incorporates all the knowledge that we have about the real population, that is, the knowledge from the sample, and it is easy to understand.

Disadvantage: The hypothetical population is not smooth, and it is probably missing information at the extremes, beyond the limits of the sample.

An alternative to this basic bootstrap is the *parametric bootstrap*. With the parametric bootstrap, we create a hypothetical population by generating random values from a theoretical distribution, based on the information in the sample. For example, we might generate lots of random values from a Normal distribution whose mean and variance can be estimated from the sample.

Definition: **The parametric bootstrap**

In the parametric bootstrap, we resample or simulate values from a theoretical distribution to observe the distribution of a sample statistic. Typically, we might simulate values from a Normal distribution that has the same mean and standard deviation as an observed sample.

Advantage: The hypothetical population contains key information from the sample, such as mean or variance, it is smooth and it contains extreme values in the tails of the distribution.

Disadvantage: If the real population is not Normally distributed, we would introduce additional error-bias-into our estimates.

RESAMPLING PROCEDURE—PARAMETRIC BOOTSTRAP

Let us illustrate the parametric bootstrap for the Toyota data.

1. Find the mean and standard deviation of the sample. Use the $n - 1$ formula for the standard deviation as it is the best estimate of the population standard deviation.
2. Using an Excel or other random number generator (RNG)*, generate 20 values from a Normal distribution with a mean and standard deviation equal to those of the sample (see Figure 7.8).
3. Record the mean of the 20 values (Figure 7.8).
4. Repeat steps two and three 1000 times.
5. Arrange the 1000 resampled means in descending order and identify the fifth and the 95th percentiles—the values that enclose 90% of the resampled means. These are the endpoints of a 90% confidence interval, as shown in Figure 7.9.

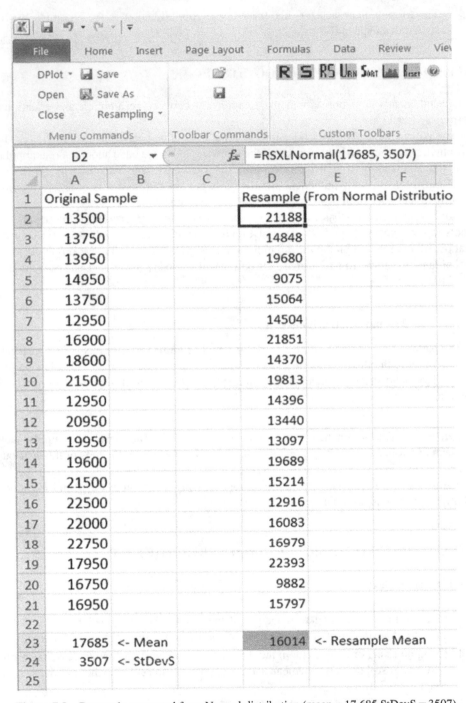

Figure 7.8   Resample generated from Normal distribution (mean = 17,685 StDevS = 3507).

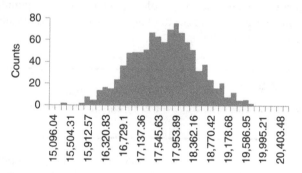

File	Home	Insert	Page Layout	Formulas	Data	Review

Open Save As

Close Resampling ▾

Save

Menu Commands   Toolbar Commands   Custom Toolbars

A2   ▾   ƒx   14820.8869688231

	A	B	C	D	E	F	G
1	1000 Simulation Results			Cell A52	16393	(5th Percentile)	
2	14821			Cell A950	19071	(95th Percentile)	
3	15218						
4	15332						
5	15687			90% Confidence Interval			
6	15733			16393 to 19071			
7	15761						
8	15801						
9	15822						
10	15835						
11	15871						
12	15896						
13	15905						
14	15907						
15	15928						

Figure 7.9   90% confidence interval, parametric bootstrap.

Figure 7.10   Histogram, parametric bootstrap (*x*-axis is resample mean).

6. In addition, Figure 7.10 shows a histogram of the 1000 resampled means and visually illustrates this confidence interval range.

Reminder: An RNG is a computer algorithm that produces numbers that are effectively random—the computer equivalent of drawing cards at random from a hat or box. An RNG that produces random numbers from a Normal distribution is such an algorithm working with a hat or box that has in it a huge supply of Normally distributed numbers written on cards. The number generation is actually a bit more involved as the numbers must be drawn from a continuous scale of real numbers. However, the computer algorithm can handle that complication and produce numbers for which a frequency distribution would be Normally shaped.

FORMULAS AND THE PARAMETRIC BOOTSTRAP

Throughout this book, we present formula counterparts to resampling procedures. These formulas calculate probabilities based on a sample mean and standard deviation and an assumption of Normality. You can see that the parametric bootstrap is very similar. It produces sampling distributions based on the mean and sample standard deviation and draws random numbers from a Normal distribution. The only difference is that the bootstrap actually simulates resample values, whereas a formula calculates where an observed value falls with respect to a standard reference distribution. See Figure 7.11 for a schematic illustration of how the simple bootstrap, the parametric bootstrap, and standard formulas relate to one another.

The first two methods in the figures "Basic Bootstrap—Theory" and "Basic Bootstrap—Practice" are functionally equivalent. The methods differ with respect to

- how they represent the hypothetical population, and
- how they determine the distribution of samples, that is, via simulation or formula.

The parametric bootstrap is sort of a hybrid. It generates simulated samples like the basic bootstrap, but like standard formulas, it uses the mean and standard deviation summary statistics to represent the hypothetical population.

APPENDIX B: ALTERNATIVE POPULATIONS

In Section 6.2, we created a simulated population from our best guess about the population, which was the sample result itself: 72 favorable to Obama, 128 unfavorable. The resulting simulation tells us about the behavior of samples drawn from a large population that is 36% positive. But, of course, we do not know that the population is 36% positive, we just used this as a good guess for our simulation.

Try It Yourself 7.5

Do the above simulation on the computer and try varying the percent positive in the box. What do you find?

Conclusion: The length of the interval produced by the best-guess population is about the same as for other plausible (nearby) populations, so it is reasonable to use the sample itself as a basis for the simulations.

APPENDIX C: BINOMIAL FORMULA PROCEDURE

Here, we show the formula-based procedure to find a confidence interval for the difference between two proportions. This procedure uses the Normal approximation to the binomial. As with other formula procedures, it is unlikely that you will need to use this as you would be getting this result via software.

Let us denote the sample sizes of the two samples as n_1 and n_2 and the proportions in the two samples as p_1 and p_2.

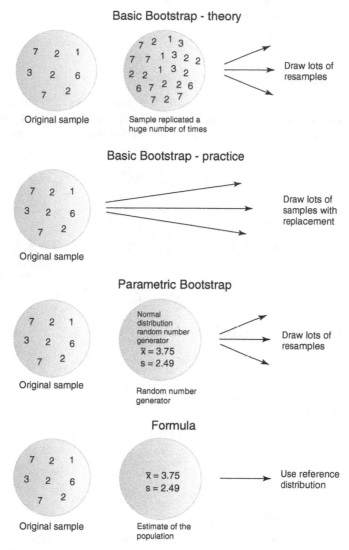

Figure 7.11 Different ways to determine how samples are distributed when the original sample consists of the values 7, 2, 1, 3, 2, 6, 7, 2.

Then the $100*(1-\alpha)\%$ z-interval for difference between population proportions is given by

$$\left(p_1 - p_2 - z_{\alpha/2} * \sqrt{\frac{p_1\left(1 - p_1\right)}{n_1} + \frac{p_2(1 - p_2)}{n_2}}, \right.$$

$$\left. p_1 - p_2 + z_{\alpha/2} * \sqrt{\frac{p_1\left(1 - p_1\right)}{n_1} + \frac{p_2(1 - p_2)}{n_2}} \right).$$

where $-Z_{\alpha/2}$ and $Z_{\alpha/2}$ are the $\alpha/2$ percentile and the $(1 - \alpha/2)$ percentile, respectively, where Z follows a standard Normal distribution.

The Binomial Formula (For Those Interested)

Suppose that n is the number of independent trials, where each trial results in either a one or a zero—success or failure. Such trials are denoted as Bernoulli trials. Let p be the probability of getting a one or success in each trial. The probability of having x ones or successes in n trials is

$$P(x) = \binom{n}{x} p^x(1 - p)^{(n-x)}.$$

x can take integer values $0, 1, 2, \ldots, n$.

The expression $\binom{n}{x}$ is called the *binomial coefficient* and is defined as follows:

$$\binom{n}{x} = \frac{n!}{x!(n - x)!}$$

The symbol ! means factorial and is defined as follows:

$$n! = n(n - 1)(n - 2) \cdots 1$$

So, the binomial coefficient $\binom{3}{1}$ is

$$\frac{3!}{1!2!} = \frac{3 \cdot 2 \cdot 1}{(1) \cdot (2 \cdot 1)} = 3$$

Binomial Formula Example

What is the probability of three successes in five trials, where the probability of a success is 0.3? So,

$$p = 0.3, \quad n = 5 \quad \text{and} \quad x = 3$$

First, the binomial coefficient

$$\binom{5}{3} = \frac{5!}{3!(5 - 3)!} = 10$$

Then, the binomial formula

$$\binom{n}{x} p^x(1 - p)^{(n-x)}$$

$$10 \cdot 0.3^3(1 - 0.3)^{(5-3)} = 10 \cdot 0.027(0.7)^2 = 0.1323$$

Normal Approximation to the Binomial

For large n, the binomial distribution can also be approximated using the Normal distribution. Then, the above calculations, using the binomial formula, need not be done. Instead, the confidence interval can be obtained using the following procedure.

If p is the point estimate of a single proportion, the $100(1 - \alpha)\%$ confidence interval for p is

$$= \left(p - z_{\alpha/2} {}^* \sqrt{\frac{p(1-p)}{n}} \, , \, p + z_{\alpha/2} {}^* \sqrt{\frac{p(1-p)}{n}} \right),$$

where n is the sample size, p is the sample proportion of ones and $-Z_{\alpha/2}$ and $Z_{\alpha/2}$ are the z-values corresponding to the $(\alpha)/2$ percentile and the $(1 - \alpha/2)$ percentile, respectively.

If we use percentage instead of proportion, then the confidence interval becomes

$$= \left(p - z_{\alpha/2} {}^* \sqrt{\frac{p(100-p)}{n}} \, , \, p + z_{\alpha/2} {}^* \sqrt{\frac{p(100-p)}{n}} \right).$$

Thus, the 90% confidence interval for p (when used as proportion) is

$$= \left(p - z_{.05} {}^* \sqrt{\frac{p(1-p)}{n}} \, , \, p + z_{.05} {}^* \sqrt{\frac{p(1-p)}{n}} \right)$$

$$\left(p - 1.645 {}^* \sqrt{\frac{p(1-p)}{n}} \, , \, p + 1.645 {}^* \sqrt{\frac{p(1-p)}{n}} \right)$$

7.10 EXERCISES

1. When you take a sample and calculate a statistic, you learn something about a population—you get an estimate of some parameter.
   (a) What is the point of then doing a resampling simulation? What additional information do you get?
   (b) What about using a formula for a confidence interval? How does its *purpose* differ from using a bootstrap simulation?

2. This problem concerns the duration of US major league baseball games. In the early 2000s, Major League Baseball was concerned about the declining television ratings of baseball. One factor considered was the length of games—did they last too long?

   For fans of other sports, here is some background relevant to this problem. Baseball has no time limit—the two teams alternate at-bats for a total of nine innings (each team's at-bat is a half-inning). If the two teams are tied at the end of nine innings, the two teams continue playing until one team is ahead at the end of a full inning. In the National League, pitchers (who are poor hitters) are required to bat. In the American League, the pitcher is not required to hit, and the pitcher's position in the batting order is taken by a designated hitter.

   Here are the durations (hours:minutes) of games played on September 13, 2011:

```
2:29 (N)
2:33 (N)
2:53 (N)
3:02 (A)
2:56 (N)
3:23 (A)
3:07 (A)
3:50 (N) (11 innings)
2:20 (A)
2:41 (A)
3:08 (N)
```

(a) Based on this sample, derive a point estimate for the AVERAGE duration of baseball games.

(b) Derive a confidence interval around that point estimate; you can use either resampling (bootstrap) or formula (by hand or software).

(c) Calculate two separate point estimates, one for National League games and the other for American League games.

(d) Comment on the difference between the two and the possible reason(s).

(e) What is the population that you are making an inference to, in part b? Is it reasonable to assume that it is Normally shaped?

(f) Is this a random sample? Comment on the validity of the sampling method.

3. A lumber yard tests the thickness of plywood panels with a sample of panels. The average thickness in the sample is 0.509 inches and the confidence interval for that mean (based on the sample) is from 0.489 inches to 0.521 inches. Which of the following is true?

(a) The observed value is within the interval; therefore, the result is statistically significant.

(b) The observed value is within the interval; therefore, the result is not statistically significant.

(c) The fact that the observed value lies within the interval is to be expected—this is how confidence intervals work.

4. Have you ever wondered, when you weigh yourself on a scale, how the scale is calibrated? How is it known that 150 pounds on that scale is the same as 150 pounds on another? Scales are calibrated by comparison to standard weights. Of course, millions of instruments require calibration and they cannot all be compared to the same weight. Rather, there is a single definitive weight and a hierarchy of other weights that are compared to it. The US National Bureau of Standards maintains a set of standard weights that are an important link in this chain.

In *Statistics*, Freedman et al. report the results of a set of definitive measurements by the Bureau of *one* 10 g standard weight, done sometime in 1962–1963. The first measurement was 9.999591. The error is very slight, on the order of what a grain of salt weights. Several other measurements also came in just slightly below 10 g. To make interpretation easier, the Bureau chose to measure not "grams," but rather "micrograms below 10 g." A microgram is one-millionth of a gram. So, instead of 9.999591, the measurement was 409. Here is a sample of 20 such measurements, all of this same weight:

409, 400, 406, 399, 402, 406, 401, 403, 401, 403, 398, 403, 407, 402, 401, 399, 400, 401, 405, 402

(a) Based on this sample, you would estimate that this standard weight falls short by _____ micrograms.

(b) The variation between one measurement and another is probably due to _____.

(c) Do a bootstrap simulation to determine how much variability there might be in your estimate in part (a). Produce a confidence interval. (This should be done on computer. If you have so far been totally unsuccessful in getting any of the software programs to work, you may do 10 simulations with actual slips of paper and a hat, report the results, and describe how you would use the results of 1000 such simulations).

(d) (OPTIONAL) Calculate this interval using statistical software and a standard formula approach.

5. The data Streams.xls reflects the pH readings on a sample of streams in a county.

(a) Calculate a point estimate for the mean pH and a confidence interval.

(b) Discuss briefly a possible sampling scheme to obtain these data and potential challenges and pitfalls.

6. Using the PulseNew.xls data:

(a) Calculate the proportion of the total who smoke and find a confidence interval around this proportion.

(b) It appears that males are more likely to be smokers than females. Find the difference in proportion who smoke (males–females) and find a confidence interval (resampling or formula) around that difference.

(c) Medical theory suggests that smoking elevates the pulse rate. Looking at the "before" pulse rate for everyone, calculate the difference: "mean smoker pulse rate" minus "mean non-smoker pulse rate". Find a confidence interval (resampling or formula) around this difference.

Answers to Questions

7.1 4/20 or 0.20 (20%).

7.2 We will review a resampling approach, which you have seen before, then introduce the binomial formula.

7.3 We are assuming that the probability of success is independent from one event to another. For example, that getting heads or tails on a coin flip does not depend on what you got on a previous flip. Or that the probability of a product being returned does not depend on whether the previous product was returned.

7.5 (the multiplication rule ...):

1. The probability of no success on five successive trials is $(0.7)^5 = 0.168$.
2. The probability of success on five successive trials is $(0.3)^5 = 0.002$.

7.6 There is no purely statistical answer to this question, at least based on this information. You would need information on the pricing and other commercial terms to make an optimal overall decision. However, the methods we have been discussing can help us determine how much confidence to place in these data—how variable they might be due to chance. We can supplement the point estimate of 0.5 oz. with a confidence interval, which is discussed next.

7.7 The original real world comparison was A minus B, so we want to parallel that in the resampling world. If the original comparison was B minus A, then the resampling world would also do the subtraction that way.

7.8 (Why draw from two boxes ...):

Drawing two resamples from two boxes answers the question "how different might this result (the 1/2 ounce advantage for vendor A) be if we had gotten different samples from each vendor?"

Drawing two resamples from one-box answers the question "if there is no difference between the vendors, how likely is it that we might get such an advantage for A, by chance?"

The two approaches are related but come from different perspectives: The two box approach is a confidence interval and regards the two vendors as different, the question is just how different? The one-box approach is a hypothesis test, regards the two vendors initially as equivalent, and seeks to disprove that assumption.

8

HYPOTHESIS TESTS

In this chapter, we move from confidence intervals to hypothesis tests, which are used in *research community* mainly because peer reviewers and regulators typically require evidence of statistical significance prior to publishing or approving results. *Data scientists* will encounter them in what marketers call A–B testing. After completing this chapter, you should be able to

- describe the difference between a hypothesis test and a confidence interval,
- correctly use the vocabulary of hypothesis testing,
- test hypotheses for
  - a single proportion
  - two proportions
  - two means
- distinguish when to use one-way and two-way hypothesis tests.

8.1 REVIEW OF TERMINOLOGY

Previously, we introduced these two concepts:

1. A study or an experiment to compare one treatment with another.
2. Effective design of sample surveys to accurately reveal characteristics of populations.

We also saw that variation comes from two random sources:

1. Which subjects get assigned to which treatments, and
2. Which individuals or elements get selected for the sample.

Introductory Statistics and Analytics: A Resampling Perspective, First Edition. Peter C. Bruce.
© 2015 John Wiley & Sons, Inc. Published 2015 by John Wiley & Sons, Inc.

We saw that resampling methods, or the corresponding formulas, could be used to quantify this random variability.

Definition: **Statistical inference**

Statistical inference is the process of accounting for random variation in data as you draw conclusions.

Confidence Intervals and Hypothesis Tests

Inference procedures, which are designed to protect you from being fooled by random variation, fall into two categories.

Confidence Intervals   Confidence intervals answer the question, "How much chance error might there be in this measurement or estimate or model, owing to the luck of the draw in who/what gets selected in a sample?"

Hypothesis Tests   Hypothesis tests answer the question, "Might this apparently interesting result have happened by chance, owing to the luck of the draw in who/what gets selected in sample(s) or assigned to different treatments?"

Essential to a hypothesis test is the concept of an imaginary null model of "nothing other than chance is happening." In our box models, we drew resamples to see if we could get outcomes as extreme as the observed outcome.

We saw that resampling methods, or the corresponding formulas, could be used to quantify this random variability.

Definition: **Null model**

A null model is an imaginary chance model (e.g., box with slips of paper) representing the idea that nothing new or novel is going on or that there is no difference between treatments A and B.

Let us look at an example from the online retail world.

Example

An online merchant has historically experienced a 10% return rate in the "kitchen gadget" category. In an effort to reduce returns, the merchant does a pilot in which it adds additional explanatory information and pictures about several products to its web site. Out of the next 200 purchases, 16 are returned. Is the pilot effective?

What is in the box: One one and nine zeros, representing the null model of 10% returns. This is different from the 8% returns we got in the sample. It is the historical rate, reflecting the "nothing other than chance is happening" hypothesis.

What we draw: 200 values with replacement.

The results: 16 or fewer returns are not unusual. See Figure 8.1.

Conclusion: A reduction in the return rate to 8% might have occurred by chance in a sample of 200. A precise count is not possible just from the histogram, but by examining

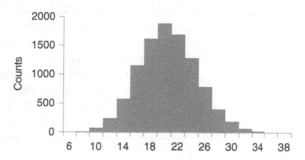

Figure 8.1   Draws, with $n = 200$, from a box with one 1 and nine 0s, x-axis is the number of 1s.

the results (not shown earlier), we determined that 219 out of 1000 simulations produced 16 or fewer ones. You can try this experiment yourself by looking at the software procedures in the textbook supplements.

p-Value

The p-value is the frequency with which a result as extreme as the observed result occurs just by chance, drawing from the null hypothesis model. In the example mentioned earlier, the p-value is 0.219.

Significance or Alpha Level

How unusual is too unusual to be ascribed to chance?

First, it is important to know that there is no natural or magic answer to this question.

Most would agree that if you observe, in the real world, an apparently "extreme" outcome, but something that extreme happens 20% or 30% of the time under a chance model, that is not rare enough to rule out chance as a possible cause.

Similarly, most would agree that if you get such an extreme result only 1% of the time under a chance model, then chance is not a likely explanation for what you saw.

What about the territory in between? Custom and tradition typically set the threshold level of statistical significance as 5%. Given an apparently extreme result seen in real life, if the chance model produces such an "extreme" result less than 5% of the time, in other words, in 5% of the resampling trials, it is said to be statistically significant. This 5% level dates from nearly a century ago from the writings of the great statistician, R. A. Fisher. He spoke of a frequency of one in 20 as being too rare to ascribe to chance.

This threshold level is termed "*alpha*" and is denoted by the Greek letter α. Alpha is determined before a study is done, usually by regulators or journal editors. As noted, it is typically set at 0.05.

Definition: Significance level

The significance level α is a threshold probability level set before performing a hypothesis test. If the hypothesis test yields a p-value equal to or less than α, the result is deemed statistically significant.

Some contend that the quest for statistical significance in the research community has gotten out of control. For some journals, a statistically significant result seems to be a necessary and sufficient condition for the publication of a study. However, statistical significance does not guarantee that the study is well designed, that the subject being studied is important, or that the result is scientifically meaningful. In the data science community, statistical significance has information value, but is not of critical decision importance.

Critical Value

Using the resampling distribution, you can determine what value of the test statistic corresponds to a given alpha level. For alpha = 0.05, for example, you would find the value of the resampled test statistic that corresponds to the 95th percentile (1−alpha = 0.95). This is termed the *critical value*. In the formula approach, you would be calculating the critical value for a standardized test statistic such as a *t*-statistic.

Question 8.1

A psychologist conducted an experiment with two groups of subjects to explore the power of suggestion and arbitrary reference points. The subjects in both groups were asked to write down their maximum bid for an item on eBay. First, though, the subjects in the treatment group were asked to look up the weather report for the next day and to note the forecast temperature. The resulting differences between the two groups appeared to suggest that the treatment did, in fact, have an effect on the bid price—focusing on the temperature even though it was irrelevant, skewed the bid price. The alpha level was 0.05 and the p-value was 0.07.

1. Is this a hypothesis test or a confidence interval question?

2. Explain briefly what the p-value means.

3. State a conclusion, incorporating both the alpha level and the p-value.

8.2   A–B TESTS: THE TWO SAMPLE COMPARISON

The two-sample comparison is a fundamental inference procedure in statistics. Is treatment A different from treatment B? We use the word "treatment" broadly—it can mean a drug, a medical device, an advertising campaign, a price, a manufacturing procedure, and so on.

If an experiment or study shows exactly identical results for A and B, the case is closed—the study shows no evidence of a difference. This rarely happens, though. More often, the results are different. The difference could be due to chance or it could be due to a real difference in the treatments.

We worry about the role of chance in fooling us into thinking that there might be a real difference. This is because we know that the human mind under-estimates the role of chance and ascribes meaning to occurrences that are merely the result of chance.

There is no foolproof way to know for certain whether an interesting result is real or is the result of chance. Hypothesis testing tells us *how* extreme the observed result is compared to what chance might produce, and it helps us make decisions on that basis.

Basic Two-Sample Hypothesis Test Concept

1. Establish a null model, which is also called *the null hypothesis*. This represents a world in which nothing unusual is happening except by chance. Usually, this null model is that the two samples come from the same population.
2. Examine pairs of resamples drawn repeatedly from the null model to see how much they differ from one another. Alternatively, we can use formulas to learn about this distribution of sample differences. If the observed difference is rarely encountered in this chance model, we are prepared to say that chance is not responsible.

Basic Two-Sample Hypothesis Test Details

1. Make sure you clearly understand
   - the sizes of the two original samples,
   - the statistic used to measure the difference between sample A and sample B, for example, the difference in means, proportions, and ratio of proportions
   - the value of that statistic for the original two samples.
2. Create an imaginary box that represents the null model. Examples could be a box with eight red chips and two black chips (to represent something with a 20% probability) or a box with all the observed body weights (in a study of how something affects body weight), where each weight is written on a slip of paper.
3. Draw out two resamples of the same size as the original samples. This can be done with or without replacement. The two procedures yield similar results, but do diverge for very small samples (<10), and both are used. The distinction between them is technical and beyond the scope of this course.
4. Record the value of the statistic of interest.
5. Repeat steps 3 and 4 many times for 1000 trials. Even more trials can be conducted for greater accuracy.
6. Note the proportion of trials that yields a value for the statistic as large as that observed.

Formula-Based Approaches

The earlier-mentioned resampling procedure is a one-size-fits-all method for two-sample comparison. Formula-based procedures are both more limited in scope and more complex. For completeness, however, we outline some of them in the following sections. Others are more advanced and are not presented here, in part because the availability of resampling alternatives has diminished the need for them.

Before getting into the formulas, you should practice with the following questions to gain a better understanding of the concept of hypothesis testing.

Practice

Question 8.2

A vehicle rental firm wants to know whether rotating tires every 12 months, compared to no rotation, increases tire lifetime. At the end of the study, the firm has tire lifetimes in months for 35 "rotate" vehicles and 29 "no rotate" vehicles. Describe a practical representation of the null model.

Question 8.3

Banner ad A generated a click-through rate of 1.2% in an experiment and banner ad B generated a click-through rate of 2.8%. How would you define the null model?

Question 8.4

An online retailer experimented with two randomly generated prices for its tablet product. Five hundred viewers saw a price of $239 and 12 purchased, whereas 720 viewers saw a price of $199 and 21 purchased. What is the null model and what do you draw from the box?

8.3   COMPARING TWO MEANS

Internal bleeding can be a serious medical problem and there is a lack in the range of remedies available for external bleeding. Internal clotting agents are a valuable treatment in such cases. Before they can be brought to market, drugs must first be proven to be effective and safe in a series of experiments, which typically start with animal studies.

Chernick and Friis (*Introductory Biostatistics for the Health Sciences*, Wiley, Hoboken, 2003, pp. 163–164) describe one such study with pigs. Ten pigs were randomly assigned to a treatment group with a new clotting agent and 10 pigs were assigned to a control group that did not receive the clotting agent. Each pig's liver was injured in a specified controlled manner and the blood loss was measured, as shown in Table 8.1.

TABLE 8.1   Blood Loss in Pigs (ml)

Control Group	Treatment Group
786	543
375	666
4446	455
2886	823
478	1716
587	797
434	2828
4764	1251
3281	702
3837	1078
Mean: 2187	Mean: 1086

Difference in the means of treatment minus control = −1101.

The null hypothesis is that the treatment has no effect and there is no difference in the average blood loss between the two groups except for what chance might produce.

Resampling Procedure

1. Write blood loss from each pig on a slip of paper and put all 20 slips of paper in a box.
2. Shuffle the box and randomly draw with or without replacement two resamples of 10 each.
3. Find the mean of each resample, subtract the mean of the first resample from the mean of the second, and record this difference.
4. Repeat steps two and three 1000 times and find out how often the recorded difference is ≤−1101 ml.

Figure 8.2 illustrates the result of 1000 repetitions of the resampling procedure. In this instance, a difference in blood loss ≤−1101 ml occurred only 47 times under the chance model. Therefore, we conclude that chance is not likely responsible and that the treatment is effective. In technical terms, the p-value is 0.047, which is less than 0.05, so the result is statistically significant.

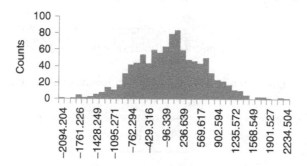

Figure 8.2   Difference in average blood loss.

8.4   COMPARING TWO PROPORTIONS

Consider again the heart attack data presented earlier in Table 7.2. The data table is repeated in Table 8.2.

The difference in proportions: $0.0741 - 0.0447 = 0.0294$.

TABLE 8.2   Cholesterol and Myocardial Infarctions (MI)

Cholesterol Level	MI	No MI	Total	Proportion
High	10	125	135	0.0741
Low	21	449	470	0.0447
Total	31	574	605	0.0512

How big is this difference compared to what chance might produce? The null hypothesis is that both high and low cholesterol groups share the same heart attack rate (come from the same population) and that such divergent samples arose by chance. The researcher would like to disprove this hypothesis. The null model that represents this hypothesis is a combined heart attack rate of 31/605 or 0.0512.

Question 8.5

In the previous chapter, we carried out a confidence interval approach for these same data. How do the two procedures differ?

CAUTION: You will notice in the resampling procedures that it does not matter whether the original group 1 is of the same size as original group 2; what is important is to use the same sample size in the resampling as were used in real life. Conventional formulas are not quite so flexible—procedures may differ, depending on whether observed sample sizes are equal or not.

Resampling Procedure

Box Sampler is limited to a sample size of 200, so this simulation must be done using Resampling Stats for Excel or StatCrunch.

1. Put 605 slips of paper in a box, 31 marked one (1) and 574 marked zero (0).
2. Shuffle the box, draw a sample of 135 with or without replacement, and find the proportion of ones.
3. Draw a second sample of 470 with or without replacement and find the proportion of ones.
4. Record the difference of the proportion from step 2 minus the proportion from step 3.
5. Repeat steps two through four 1000 times. How often was the difference ≥ 0.0294?

Figure 8.3 displays a histogram of the results of the resampling procedure. Out of all 1000 resamples, only 74 had a difference ≥ 0.0294 for a *p*-value of 0.074. This falls just above the usual standard for statistical significance, 0.05, so we conclude that chance might be responsible and the result is not statistically significant. It lies close to being statistically

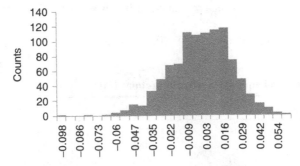

Figure 8.3   Difference in proportions.

significant, however, and subsequent studies confirmed the connection between cholesterol and heart disease.

The statistical literature contains resampling/permutation procedures for multiple sample testing both with and without replacement—more often without replacement. The difference in their statistical properties is beyond the scope of this course.

Specific software procedures for these examples are found in the corresponding sections of the textbook supplements.

8.5   FORMULA-BASED ALTERNATIVE—*t*-TEST FOR MEANS

The resampling approach described earlier and the formula approaches that follow are alternates. Either may be used because they are roughly equivalent. The resampling procedure is fundamentally the same and conceptually simple—what is in the box(es), how many resamples should be drawn, and what sizes should they be. The formula approaches are more complex and originated from the days when resampling was unavailable.

The basic *t*-test procedure for a difference in means—two-sample comparison—requires sample sizes to be large enough such that the difference in means is roughly Normally distributed. How large is "large enough" depends on the data—data that are extremely skewed require larger sample sizes. For data that are reasonably Normally distributed themselves, the *t*-test can be used when the sample sizes exceed 15. Here is the procedure.

The null hypothesis is that the two samples come from the same population.

Let us denote the sample sizes of the two samples as n_1 and n_2, the sample means as $\overline{X}_1$ and $\overline{X}_2$, and the sample variances as s_1^2 and s_2^2.

As we are testing the hypothesis that the two samples could come from the same population, we can calculate the null model variance by pooling the two samples together. This is known as "*pooled variances*."

Then, the sample statistic is given by

$$t = \frac{\overline{X}_1 - \overline{X}_2}{S\sqrt{\frac{1}{n_1} + \frac{1}{n_2}}}$$

where

$$S^2 = \frac{(n_1 - 1)s_1^2 + (n_2 - 1)s_2^2}{n_1 + n_2 - 2}.$$

t follows Student's t distribution with $n_1 + n_2 - 2$ degrees of freedom.

$-t_{n_1+n_2-2,\,\alpha/2}$ and $t_{n_1+n_2-2,\,\alpha/2}$ are the *t*-values corresponding to the $\alpha/2$ percentile and the $(1 - \alpha/2)$ percentile.

Using the sample data, we calculate the value of the sample statistic *t*.

If the calculated *t*-value is less than $-t_{n_1+n_2-2,\,\alpha/2}$ or greater than $t_{n_1+n_2-2,\,\alpha/2}$, we reject the null hypothesis and conclude at significance level α that the two population means are not equal. In other words, the two samples are drawn from two different populations rather than the same population.

If $-t_{n_1+n_2-2,\,\alpha/2} < t < t_{n_1+n_2-2,\,\alpha/2}$, then we do not reject the null hypothesis. We conclude that the samples may come from the same population and that the population means may be the same.

8.6   THE NULL AND ALTERNATIVE HYPOTHESES

In hypothesis testing, our goal has been to determine whether an apparent effect is real or might be due to chance. A key part of the procedure is to formulate a null hypothesis, or null model, that embodies the idea that chance is responsible. The null hypothesis is a straw man—if you can disprove it, you have made progress.

Formulating the Null Hypothesis

It is a good idea to have both a conceptual and a firm quantitative idea of the null model. Below, we list some examples of possible null hypotheses that you would like to be able to disprove. We describe their conceptual form, provide a quantitative translation, and outline an implied box model.

1. *Null hypothesis:* Advertisements A and B are equally good. We want to disprove the null hypothesis so we can identify an advertisement that is definitively better. *Possible quantitative translation:* Advertisements A and B both have the same click-through rate. *Possible box model:* A single box with 0s—no clicks—and 1s—clicks—where the 1s are the total clicks for (A plus B) and the 0s are the total page-views.

2. *Null hypothesis:* No-fault reporting in hospitals is no more effective than the regular system. We hope to disprove the null hypothesis to establish that the new no-fault system is better. *Possible quantitative translation:* The no-fault system and the regular system will both reduce errors to the same degree. *Possible box model:* A single box with the total number of errors for both groups.

3. *Null hypothesis:* A new, targeted chemotherapy for advanced breast cancer is no more effective than the standard tamoxifen. We want to disprove the null hypothesis so that we can prove that the new therapy is better. *Possible quantitative translation:* Median survival time in a clinical trial is the same for both groups. *Possible box model:* A single box with a slip for each patient (regardless of the group), with the number of days the person survived on the slip.

4. *Null hypothesis:* A new and costly manufacturing process will not increase the chip-processing speed sufficiently to be worth the investment. We want to disprove the null hypothesis so that we know definitively that the new process produces faster chips. *Possible quantitative translation:* Mean processing speed in a pilot new-process sample falls short of a 25% improvement. *Possible box model:* A single box with a slip for each sample chip tested—both new-process and existing process—with the processing speed written on the slip.

Corresponding Alternative Hypotheses

The alternative hypothesis is the theory you would like to accept, assuming that your results disprove the null hypothesis. Below, we present some alternative hypotheses corresponding to the earlier described null models.

1. Advertisements A and B are different. In this case, we do not care which is better. We just want to know if one is better to a statistically significant degree so that we can feel confident using it.

2. The no-fault system is an improvement over the regular system, that is, it reduces errors more.

3. The new, targeted chemotherapy is better than tamoxifen.

4. The new manufacturing process improves chip speed by at least 25%.

One-Way or Two-Way Hypothesis Tests

How we formulate the alternative hypothesis determines whether we perform a one-way or a two-way hypothesis test.

Consider the no-fault error reporting system introduced in Canadian hospitals to determine whether the apparent reduction in error rates might have been due to chance (Table 8.3).

TABLE 8.3   Average Error Reduction in Hospitals

Control group	1.88
No-fault reporting group	2.80
Difference	0.92

In the resampling simulation that we carried out earlier in this text, we asked whether a *difference* of 0.92 in error reduction might have come about by chance. What we really want to know is whether an *improvement* of 0.92—no-fault over control—might have happened by chance. Figure 8.4 again shows the distribution of chance results, in which we repeatedly shuffled the hospital error reduction counts by combining control and no-fault in a box and drawing pairs of resamples of 25.

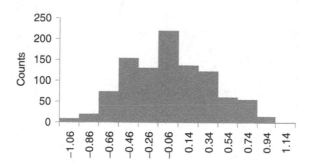

Figure 8.4   Permuted hospital error reductions—resample 1 mean minus resample 2 mean.

Should we count the extreme results at each end—≥ 0.92 and ≤ -0.92? Or should we count only the extreme results at the positive end—≥ 0.92?

The Rule

If the alternative hypothesis is unidirectional, perform a one-way or a one-tailed test. In other words, if the original study requires a difference in a specific direction, count only those differences. An example of a unidirectional—one-way—hypothesis is the question of whether a treatment is *better* than the control, as illustrated in Figure 8.5.

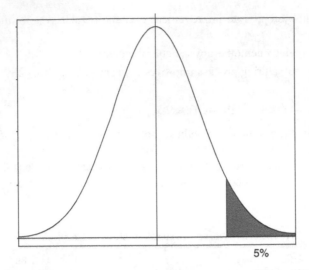

Figure 8.5   One-tailed test—this Illustration is for Alpha = 0.05.

If the alternative hypothesis is bi-directional, then the differences in either direction should be counted. Such a two-way test is exemplified by the question of whether advertisement A or B is better. The two-way test is illustrated in Figure 8.6

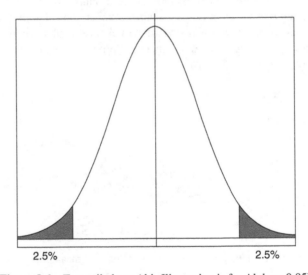

Figure 8.6   Two-tailed test (this Illustration is for Alpha = 0.05).

The Why

Recall that the purpose of the hypothesis test is to keep you from being fooled by chance events into thinking that something interesting is happening.

If the something interesting is an apparent difference in either direction—which ad works better?—then you need to compare the observed difference with the extreme chance results in either direction.

Suppose that the something interesting is a difference in only one direction—treatment better than control. In such a case, you are not going to take action on a difference in the other direction and, therefore, you do not need protection against being fooled by chance results in that direction. You only need to compare the observed results with the extreme chance results in just one direction.

Practice

Question 8.6

Assess whether these situations call for a one-tailed or a two-tailed test.

1. *The intelligence of people is measured as a score on an IQ test. Can a specific set of drill-study exercises increase the IQ score of a person?*
2. *ABC pizzeria claims that on average, each of their large pepperoni pizzas is topped with two ounces of pepperoni. Many customers have lodged a complaint against them saying that the actual amount of pepperoni used is considerably less than that. So measurements of pepperoni weight are taken on some pizzas.*
3. *In a typical population, platelet counts are expected to range from 140 to 440 (thousands of platelets per cubic millimeter of blood). The platelet counts of a group of elderly women are observed to see if their counts are abnormal.*
4. *The pH of the catalyst in a chemical reaction is measured by a standard titration procedure. For the reaction to proceed at a desired rate, the pH of the catalyst should be close to 6.4. A chemist wishes to test whether the pH is different from 6.4 by using some readings.*
5. *Blood pressures of a group of patients were measured before and after a full meal. Does blood pressure increase, on average, after the intake of a full meal?*
6. *A manufacturer of pipe wants to test whether a special coating can retard the corrosion rate of pipes used in laying underground electrical cables. He coats some pipes with a new plastic and takes relevant readings on them after 1 year to check whether the plastic coating was beneficial.*
7. *A contractor buys cement from a manufacturer. The cement bags are supposed to weigh 94 pounds. The contractor wants to test if he is getting his money's worth, so he measures the weights of a sample of cement bags.*
8. *A drug is tested to determine whether it can lower the blood glucose level of diabetic rats. One group of rats is given the drug and the other group is used as control, which is not given the drug. The blood glucose levels of the two groups are measured to test the drug's effectiveness.*
9. *We want to test whether the heart weights of male and female cats differ significantly. Therefore, the heart weights of a group of male and female cats are measured.*
10. *The number of defectives in a group of articles from two different factories is observed. We want to test if the product of the second factory is superior to that of the first factory.*

8.7 PAIRED COMPARISONS

Much has been written about the supposed contribution of music to cognitive learning. Consider these hypothetical reading comprehension scores for 11 subjects—initially after reading a passage without background music and then a week later after reading a similar passage with background music (Table 8.4).

TABLE 8.4   Reading Scores

Subject #	Without Music	With Music
1	24	27
2	79	80
3	17	18
4	50	50
5	98	99
6	45	47
7	97	97
8	67	70
9	78	79
10	85	87
11	76	78
Mean	65.10	66.55

The observed difference in means is 1.45. Following our standard two-sample testing procedure in which the null hypothesis is that all reading scores belong to the same population, we would do the following:

1. Place all 22 values in a box.
2. Take a random resample of 11, then another resample of 11. These resamples could be with or without replacement. In this case, we will do it without replacement.
3. Find the mean of each and record the difference—second resample minus the first, which is the way we did it in the table shown earlier.
4. Repeat steps two through three 1000 times.

Figure 8.7 is a histogram of the distribution of differences.

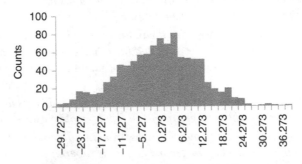

Figure 8.7   Permutation distribution of reading scores.

The observed difference of 1.45 is right in the center of the distribution, which is not at all extreme compared to what chance can produce. Thus, our hypothesis test produces the conclusion of no difference.

But let us look again at the data (Table 8.5).

TABLE 8.5   **Hypothetical Reading Scores—Paired Differences**

Subject #	Without Music	With Music	Difference
1	24	27	+3
2	79	80	+1
3	17	18	+1
4	50	50	0
5	98	99	+1
6	45	47	+2
7	97	97	0
8	67	70	+3
9	78	79	+1
10	85	87	+2
11	76	78	+2

In every case, music improved the individual's reading score or left it unchanged. In no case did the reading score worsen. Our intuition is not happy with the no-difference conclusion.

Here is what is happening. The reading score improvements are very small compared to the differences between individuals. In our standard two-sample comparison, the differences get swamped by the variation among subjects. The standard test, whether a resampling or a *t*-test, cannot discern the treatment effect in the face of all the between-subject variation. In statistical terms, it lacks power.

Definition: **Power**

Power is the probability that a statistical test will identify an effect, that is, determine that there is a statistically significant difference when one exists. To calculate power, you need to know (i) the effect size you want to discern, (ii) the sample sizes, and (iii) something about sample variances and distribution.

What we need for the reading score's example is a null hypothesis model that does not treat all subjects as belonging to the same population. Instead, we could have a null model in which each individual is in his or her own population, and the subjects are not mixed. Such a null model would be described as each individual has two scores, and it is a matter of chance which score is for music and which for no-music. We can test this null model with a paired-comparisons test.

Paired Comparisons: Resampling

1. Randomly shuffle the two scores for the first subject into columns 1 and 2.
2. Repeat the shuffle for the remaining 10 subjects.
3. Calculate the mean scores for columns 1 and 2 and record the difference, that is, column 2 minus column 1.

4. Repeat steps one through three 1000 times.
5. Draw a histogram of the resampled differences and find out how often the resampled difference exceeds the observed value of 1.45.

The histogram in Figure 8.8 shows the results.

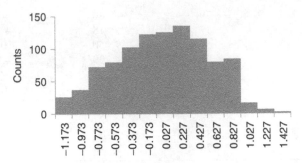

Figure 8.8   Paired resampling test.

The observed difference of 1.45 now looks extreme compared to what chance produced. Chance resulted in a difference of 1.45 very rarely—it does not even show up on the histogram. Therefore, we conclude that the difference is statistically significant.

Paired Comparisons: *t*-Test

The formula-based counterpart to this procedure is a paired *t*-test. This is how it is conducted.

The null hypothesis is that there is no difference in the two readings.

Let us denote the sample size as n, the first and second readings of the same subject as X_1 and X_2, and the differences $X_2 - X_1$ as D.

The sample statistic is given by

$$t = \frac{\overline{D}}{\frac{S_D}{\sqrt{n}}}$$

where $\overline{D}$ is the mean of the differences and S_D is the standard deviation of the differences D.

t follows Student's t distribution with $n - 1$ degrees of freedom.

This test will generally be a one-tailed test, that is, we will test whether the difference is more positive than chance would produce—such as showing improvement in the reading score with music—or more negative—such as weight reduction after following a new diet. Accordingly, we use either $t_{n-1, \alpha}$, the $(1 - \alpha)$th percentile or $-t_{n-1, \alpha}$, the αth percentile.

Using the sample data, we calculate the value of the sample statistic t.

Case 1: Testing if the Difference is Positive   If calculated $t > t_{n-1, \alpha}$, we reject the null hypothesis and conclude at significance level α that the difference between the two readings is significant.

If $t < t_{n-1,\alpha}$, then we do not reject the null hypothesis and conclude that the difference between the two readings is not significant and can be attributed to chance.

Example

For the reading scores, the observed value of $D = 1.4545$ and SD $= 1.0357$, and $n = 11$. So the calculated value of t is:

$$t = \frac{1.4545}{\frac{1.0357}{\sqrt{11}}} = 4.66$$

The critical value of t for alpha $= 0.05$ (obtained via web search for "t-table") is

$$t_{10,\ 0.05} = 1.81$$

As $4.66 > 1.81$, we conclude that the difference is statistically significant—it is not within the range of what chance might account for. This is illustrated in Figure 8.9:

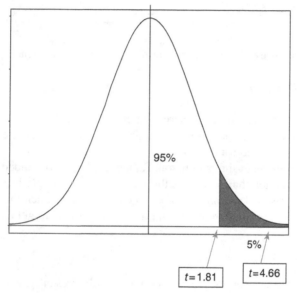

Figure 8.9   Paired t-test, reading results. The t-statistic for the observed results falls at 4.66, well beyond chance results, which are bounded by 1.81.

Case 2: Testing if the Difference is Negative   If calculated $t < -t_{n-1,\ \alpha}$, we reject the null hypothesis and conclude at significance level α that the difference between the two readings is significant.

If $t > -t_{n-1,\alpha}$, then we do not reject the null hypothesis and conclude that the difference between the two readings is not significant and can be attributed to chance (Figure 8.10).

APPENDIX A: CONFIDENCE INTERVALS VERSUS HYPOTHESIS TESTS

Immediately above, we laid out the logic of a hypothesis test. Earlier, we practiced calculating confidence intervals. Before continuing with more on hypothesis tests, let us look at the example shown earlier in the context of a confidence interval and then review the difference between the two.

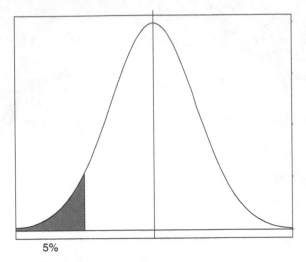

5%

Figure 8.10   Paired *t*-test, testing for a negative difference.

CONFIDENCE INTERVAL

In a confidence interval, we have a measurement or an estimate based on a sample. We want to know how inaccurate that measurement might be, based solely on the luck of the draw in who or what gets selected for a sample.

There is no null model of the population. Rather, the box in our model simply contains a hypothetical population that replicates the features of the sample itself as faithfully as possible. We draw resamples from the box to determine how much they differ from one another. From that, we can learn how much our original sample might be in error.

Example

This example uses the same online merchant sample results presented at the beginning of the chapter, but with a different scenario, to illustrate the question that the confidence interval seeks to address. An online merchant wants to get an accurate estimate of its return rate. Out of the next 200 purchases, 16 are returned. This is an 8% return rate, but how accurate is this estimate?

What is in the box: 16 ones and 184 zeros, representing the actual sample results.

What we draw: 200 values with replacement. Software procedures for this example are found in the textbook supplements.

The histogram in Figure 8.11 shows the results.

Conclusion: The histogram indicates how variable the results might be in a sample of 200 with a return rate of 8%. To get a 90% confidence interval, we would chop off 5% at either end, which in this case yields an interval of 10–22. The result would be reported in this way.

Point estimate of return rate: 8%

90 percent confidence interval: 5% returns to 11% returns.

You may also see it written as shown below with just the confidence interval indicated. The symbol π is often used in statistics to refer to a proportion.

$0.05 < \pi < 0.11$

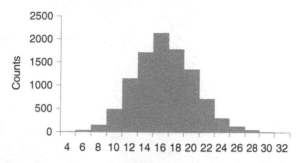

Figure 8.11   Draws, with $n = 200$, from a box with 16 1s and 184 0s, x-axis is the number of 1s.

Question 8.7

Which of the following statements is more true:

1. *Because 8% (.08) is within the confidence interval, we conclude that our estimate of the return rate is correct.*
2. *We do not know exactly what the return rate is, but it probably lies between 5% and 11%.*

RELATIONSHIP BETWEEN THE HYPOTHESIS TEST AND THE CONFIDENCE INTERVAL

Figures 8.12 and 8.13 illustrate the relationship between the hypothesis test, based on a null model, and the confidence interval, based on the sample result.

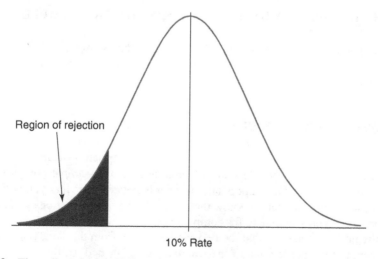

Figure 8.12   The null model is a 10% return rate. We reject the null hypothesis if the sample result falls in the rejection region.

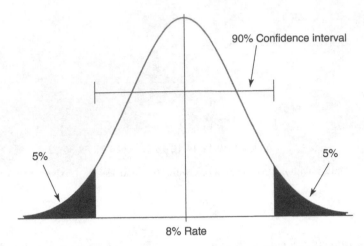

Figure 8.13 The sample estimate is an 8% return rate. The confidence interval shows how much the same estimate might be in error.

COMMENT

In the example presented at the beginning of the chapter, we tested a hypothesis. In this example, we had a 10% return rate in the box because this was the historical rate. We wanted to know whether a sample with an 8% return was unusual.

In the second example, we calculated a confidence interval. We had a sample result of 8% and we wanted to know how much it might deviate from the true rate.

Some books will use the confidence interval results to answer the hypothesis test question. The logic goes like this: Given our sample result of 8%, is it reasonable to believe that the historical rate could be 10%? In this case, as 10% is within the confidence interval for the sample result of 8%, the answer is yes.

APPENDIX B: FORMULA-BASED VARIATIONS OF TWO-SAMPLE TESTS

In software, you may encounter various alternatives to the two-sample tests outlined thus far. They are included here for reference, but they are not covered in depth for the reasons explained in the following sections.

Z-TEST WITH KNOWN POPULATION VARIANCE

When the population variance is known, a Z-test is appropriate instead of a t-test. This is a rare circumstance. The whole point of inference for two-sample comparisons is to draw conclusions about unknown populations, so it is not common that you will know the variance of the population, yet not know the answer to the other questions you have about it. The procedure is presented here for completeness.

Again the null hypothesis is that the two samples come from the same population.

Let us denote the sample sizes of the two samples as n_1 and n_2, the sample means as $\overline{X}_1$ and $\overline{X}_2$, and the known population variance as σ^2. The population standard deviation $\sigma = \sqrt{\sigma^2}$.

Now, the sample statistic is given by

$$Z = \frac{\overline{X}_1 - \overline{X}_2}{\sigma\sqrt{\frac{1}{n_1} + \frac{1}{n_2}}}$$

where Z follows a standard Normal distribution with mean $= 0$ and standard deviation $= 1$.
$-Z_{\alpha/2}$ and $Z_{\alpha/2}$ are the $\alpha/2$ percentile and the $(1 - \alpha/2)$ percentile, respectively.

Using the sample data, we calculate the value of the sample statistic Z.

If calculated $Z < -Z_{\alpha/2}$ or $Z > Z_{\alpha/2}$, we reject the null hypothesis. We conclude at significance level α that the two population means are neither equal nor are the two samples drawn from two different populations.

If $-Z_{\alpha/2} < Z < Z_{\alpha/2}$, then we do not reject the null hypothesis. We conclude that samples may come from the same population and that the population means may be the same.

POOLED VERSUS SEPARATE VARIANCES

Mathematically, it is possible to conduct a t-test either using separate variances for the two samples or using a single pooled estimate of variance. Typically, the hypothesis being tested is that the two samples come from the same population, reflecting the principle that we assume that nothing other than chance is happening to distinguish the two samples. We want to see if we can be proven wrong. Although the t-test can be performed mathematically with separate variances, using that approach would not test the same-population hypothesis. Instead, it would test the hypothesis that the two samples come from two populations with different variances but with the same mean, which is a fairly unusual circumstance.

Again, the procedure is presented here for completeness.

The null hypothesis is that the two populations, from which the two samples are drawn, have equal means, but different variances.

Again, the sample sizes of the two samples are denoted as n_1 and n_2, the sample means as $\overline{X}_1$ and $\overline{X}_2$, and the sample variances as s_1^2 and s_2^2.

Now, the sample statistic is given by

$$t = \frac{\overline{X}_1 - \overline{X}_2}{\sqrt{\frac{s_1^2}{n_1} + \frac{s_2^2}{n_2}}}$$

t follows Student's t distribution with v degrees of freedom, where

$$v = \frac{\left(\frac{s_1^2}{n_1} + \frac{s_2^2}{n_2}\right)^2}{\frac{\left(\frac{s_1^2}{n_1}\right)^2}{n_1-1} + \frac{\left(\frac{s_2^2}{n_2}\right)^2}{n_2-1}}$$

$-t_{v,\alpha/2}$ and $t_{v,\ \alpha/2}$ are the t-values corresponding to the $\alpha/2$ percentile and the $(1 - \alpha/2)$ percentile, respectively.

Using the sample data, we calculate the value of the sample statistic t.

If calculated $t < -t_{v,\ \alpha/2}$ or $t > t_{v,\ \alpha/2}$, we reject the null hypothesis. We conclude at significance level α that the two population means are not equal.

If $-t_{v,\ \alpha/2} < t < t_{v,\ \alpha/2}$, then we do not reject the null hypothesis. We conclude that the two populations, from which the two samples are drawn, may have equal means.

FORMULA-BASED ALTERNATIVE: Z-TEST FOR PROPORTIONS

In this section, we discuss the formula-based counterpart to the resampling test for proportions outlined earlier.

The null hypothesis is that the proportions in the two populations, from which the two samples are drawn, are equal.

Let us denote the sample sizes of the two samples as n_1 and n_2 and the proportions in the two samples as p_1 and p_2.

As we are testing the hypothesis such that the population proportions are equal, we can obtain the unknown population proportion by pooling the sample proportions together.

Thus, the sample statistic is given by

$$Z = \frac{p_1 - p_2}{\sqrt{\hat{p}(1 - \hat{p})\left(\frac{1}{n_1} + \frac{1}{n_2}\right)}}$$

where $\hat{p} = \frac{n_1 p_1 + n_2 p_2}{n_1 + n_2}$.

Z follows a standard Normal distribution.

$-Z_{\alpha/2}$ and $Z_{\alpha/2}$ are the $\alpha/2$ percentile and the $(1 - \alpha/2)$ percentile, respectively.

Using the sample data, we calculate the value of the sample statistic Z.

If calculated $Z < -Z_{\alpha/2}$ or $Z > Z_{\alpha/2}$, we reject the null hypothesis and we conclude at significance level α that the proportions in the two populations are not equal.

If $-Z_{\alpha/2} < Z < Z_{\alpha/2}$, then we do not reject the null hypothesis and we conclude that the proportions in the populations may be equal.

This formula relies on an approximation—when sample sizes n_1 and n_2 are large enough and/or the proportions in each sample p_1 and p_2 are close enough to 0.5, the difference between the two proportions is Normally distributed.

What is large enough? One guideline is that the following conditions must all hold true for this Normal approximation to be valid.

$$n_1 \times p_1 \geq 5$$
$$n_2 \times p_2 \geq 5$$
$$n_1 \times (1 - p_1) \geq 5$$
$$n_2 \times (1 - p_2) \geq 5$$

As you can see, low probability events require large sample sizes. A sample size of 100 would just suffice in testing a proportion of 5%. However, if the percentage drops to 1%, the required sample size jumps to 500.

The Z-test for proportions is still frequently encountered, although the resampling permutation test outlined earlier is more versatile. It requires no approximations and can operate with small samples and extreme proportions.

8.8   EXERCISES

1. Answer the following questions referring to the PulseNew.xls data:

   (a) Is there any sign that the proportion who ran is different for males versus females? Perform a hypothesis test (resampling or formula), state your conclusion in plain English that would be understandable to someone who never took a statistics course.

   (b) Is there evidence here that the "before" pulse rate differs between smokers and nonsmokers? Medical theory suggests that smoking elevates the pulse rate. Perform a hypothesis test (resampling or formula).

   (c) We would expect running in place to affect a person's pulse rate. Test whether before and after pulse rates are the same.

2. Which is a more effective marketing message—a plain text email or email pictures and design elements? Statistics.com performed a test in June 2012, sending an email to a list that was randomly split into two groups, sometimes called *a split A/B test*. For some reason, the email service used by Statistics.com did not produce an even split. Here are the results:

   Group A—plain text message: 71 sent, 13 opens
   Group B—same content, with images: 355 sent, 47 opens

   (a) The marketing manager worries that the imbalance in the sample sizes renders the experiment invalid. He thinks that, to have statistical validity, the two groups must be of roughly equal sizes. Is this correct?

   (b) Looking at the results, what is the difference between group A and B, expressed in useful and easy-to-understand units? Is this what you were expecting?

   (c) Next, we consider whether the difference between group A and group B is statistically significant. State the null hypothesis, and state whether we should perform a one-tail or two-tail hypothesis test?

   (d) Here are the results of two software procedures, one resampling and one formula-based. Assuming that the correct procedures were followed, interpret either output to answer the question of whether the difference in proportions is statistically significant.

   FORMULA RESULTS (StatCrunch 2-sample test of proportions)

   ```
   Hypothesis test results:
   p1 : proportion of successes for population 1
   p2 : proportion of successes for population 2
   p1 - p2 : difference in proportions
   H0 : p1 - p2 = 0
   HA : p1 - p2 ≠ 0
   ```

Difference	Count1	Total1	Count2	Total2	Sample Diff.	Std. Err.	Z-Stat	P-value
p1 - p2	13	71	47	355	0.0507	0.0452	1.1212	0.2622

RESAMPLING RESULTS

Procedure:

1. Combine $13 + 47 = 60$ slips of paper marked 1 (open) and $58 + 308 = 366$ slips of paper marked 0 (not opened) in an urn.

2. Shuffle the urn and draw (with replacement) one resample of 71 (resample 1) and another of 355 (resample 2).

3. Find the proportion of 1s in resample 1 and in resample 2.

4. Record the difference in proportions.

5. Repeat steps 2–4, say, 1000 times.

6. Find out how often we got a difference (in either direction—either in favor of plain text or in favor of images) as extreme as the observed difference of 0.051.

Results:

Histogram of resampled difference in proportions:

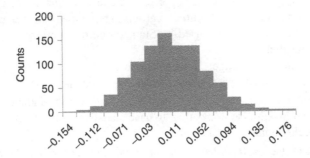

114 of the 1000 trials produced a difference in proportions ≥ 0.0507.

124 of the 1000 trials produced a difference in proportions ≤ -0.0507.

Estimated p-value: $(114 + 124)/1000 = 0.238$.

(e) The example shown earlier was simplified for presentation. The actual numbers were

Group A—plain text message: 710 sent, 130 opens

Group B—same content, with images: 3550 sent, 470 opens

a. What is the difference between group A and B, expressed in useful and easy-to-understand units?

b. To assess statistical significance, the below resampling procedure was used in the EARLIER problem in the preceding section. How would you modify it to solve this revised version of the problem? Report the modified procedure.

EARLIER PROCEDURE

1. Combine $13 + 47 = 60$ slips of paper marked 1 (open) and $58 + 308 = 366$ slips of paper marked 0 (not opened) in an urn.

2. Shuffle the urn and draw (with replacement) one resample of 71 (resample 1) and another of 355 (resample 2).

3. Find the proportion of 1s in resample 1 and in resample 2.

4. Record the difference in proportions.

5. Repeat steps 2–4, say, 1000 times.

6. Find out how often we got a difference (in either direction—either in favor of plain text or in favor of images) as extreme as the observed difference of 0.051.

3. A web-based real-estate information and valuation site wants to determine whether it will be worth it to expend time and money to implement a more sophisticated modeling system to estimate home values. A key question is whether the model is more accurate. A firm conducts an experiment in which it tries out the model and obtains valuations on 15 homes that were recently sold (group A). It also samples 15 other homes from its records that were recently sold (group B). It then compares the valuation error in the two groups. To simplify the analysis, it looks first at the absolute value of the valuation error (not caring about sign).
Valuation error (absolute value in $000)

Group A—35, 15, 17, 4, 13, 22, 19, 43, 33, 21, 5, 9, 18, 20, 3.
Group B—23, 44, 9, 34, 39, 49, 22, 22, 7, 11, 29, 17, 35, 6, 11.

Note: To make things easier to present, the sample size shown here, 15 in each group, is smaller than would normally be analyzed by a firm in this situation.

Answers to Questions

8.1 1. Hypothesis test. (We are testing whether the results might be due to chance.)

2. Even if there is no real difference between the two groups, 7% of the time you would get a difference between the two groups as big as the observed difference, just by chance.

3. Although the results appeared to suggest a difference between the two groups, the *p*-value is not low enough to meet the threshold you established (the alpha) for statistical significance.

8.2 (vehicle rental firm …):

35 slips of paper with the lifetimes for the "rotate" vehicles and 29 slips of paper with the lifetimes for the "no rotate" vehicles, all combined in a box, from which you will draw resamples of sizes 35 and 29.

8.3 (banner ads …):

We do not know how to define the null model. All we have are the rates, we are missing the sample sizes! Consider these two cases:

Situation #1:

- Ad A had three clicks in 250 views (1.2%)
- Ad B had 28 clicks in 1000 views (2.8%)

The null model = 31 clicks in 1250 views (2.48%) This null model can be expressed by any box that contains 2.48% clicks.

Situation #2:

- Ad A had 12 clicks in 1000 views (1.2%).
- Ad B had seven clicks in 250 views (2.8%).

The null model = 19 clicks in 1250 views (1.52%). This null model can be expressed by any box that contains 1.52% clicks.

8.4 (online retailer ...):
The box contains 1220 slips of paper, 33 marked "purchase," and you draw samples of sizes 500 and 720.

8.5 With the confidence interval procedure, we took the world as we saw it and drew from two separate urns. With the hypothesis test, we imagined the single-urn null model. The result from the earlier confidence interval procedure is consistent with the earlier hypothesis test —as the confidence interval included 0 and negative values, we could not rule out the possibility that the higher heart attack rate for men with high cholesterol was due to chance.

8.6 Practice questions:

1. The intelligence of people is measured as a score on an IQ test. Can a specific set of drill-study exercises increase the IQ score of a person? (one tail)

2. ABC pizzeria claims that on average, each of their large pepperoni pizzas is topped with two ounces of pepperoni. Many customers have lodged a complaint against them saying that the actual amount of pepperoni used is considerably less than that. So measurements of pepperoni weight are taken on some pizzas. (one tail)

3. In a typical population, platelet counts are expected to range from 140 to 440 (thousands of platelets per cubic millimeter of blood). The platelet counts of a group of elderly women are observed to see if their counts are abnormal. (two tail)

4. The pH of the catalyst in a chemical reaction is measured by a standard titration procedure. For the reaction to proceed at the desired rate, the pH of the catalyst should be close to 6.4. A chemist wishes to test whether the pH is different from 6.4 by taking some readings. (two tail)

5. Blood pressures of a group of patients were measured before and after a full meal. Does blood pressure increase, on average, after the intake of a full meal? (one tail)

6. A manufacturer of pipe wants to test whether a special coating can retard the corrosion rate of pipes used in laying underground electrical cables. He coats some pipes with a new plastic and takes relevant readings on them after 1 year to check whether the plastic coating is beneficial. (one tail)

7. A contractor buys cement from a manufacturer. The cement bags are supposed to weigh 94 pounds. The contractor wants to test whether he is getting his money's worth, so he measures the weights of a sample of cement bags. (one tail)

8. A drug is tested to determine whether it can lower the blood glucose level of diabetic rats. One group of rats is given the drug and the other group is used as control, which is not given the drug. The two groups' blood glucose levels are measured to test the drug's effectiveness. (one tail)

9. We want to test whether the heart weights of male and female cats differ significantly. Therefore, the heart weights of a group of male and female cats are measured. (two tail)

10. The number of defectives in a group of articles from two different factories is observed. We want to test whether the product of the second factory is superior to that of the first factory. (one tail)

8.7 #2 is more true. Bracketing our estimate with a confidence interval gives a range of possible *error* but does not add information about whether the estimate is *correct*.

9

HYPOTHESIS TESTING—2

Thus far, we have been looking mainly at comparisons between sample A and sample B. Hypothesis testing is used in other situations as well. As with the previous chapter, the material in this chapter will be of interest primarily to *researchers*; *data scientists* will have less need of it. After completing this chapter, you should be able to:

- test a hypothesis for a single proportion or mean,
- test a hypothesis involving multiple samples of count data and difference from expectation under a null model.

9.1 A SINGLE PROPORTION

In medical experiments, or studies aimed at publication, experiments with a single group are relatively uncommon—we saw earlier that well-designed experiments involve a control group and a treatment group to ensure that the only effect being evaluated is the effect of the treatment.

In business, however, single-group tests may occur to determine whether a treatment produces a change from the status-quo, particularly when a fully controlled experiment is infeasible or too costly.

Example

A sports gym offers trial memberships, and historically 25% of the trials get converted to full memberships within a week after the expiration of their trial membership. Then, the gym decides to try out a new procedure in which an employee calls the new customer

Introductory Statistics and Analytics: A Resampling Perspective, First Edition. Peter C. Bruce.
© 2015 John Wiley & Sons, Inc. Published 2015 by John Wiley & Sons, Inc.

1 week into the trial period to see how things are going, what the customer is interested in, and how the gym employees might help the new customer get more involved. After 3 months, 165 trials have been extended to potential new customers, and 53 of them have converted to full paying memberships.

In such a case, we want to determine whether a sample of 165 could produce 53 conversions, that is, 32%, if the true rate of conversion is still only 25%.

Resampling Procedure

1. Put four slips of paper into the box—one marked 1 for conversions and the other three marked zero for nonconversions.
2. Make 165 random draws from the box, with replacement, each time noting the outcome. Record the total number of ones drawn.
3. Repeat step two 1000 times, recording the number of ones drawn in each sample of 165.
4. Find out how often we got 53 or more ones.

Figure 9.1 shows the histogram of the results.

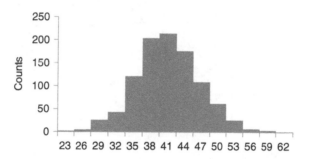

Figure 9.1   Histogram for customer conversions at gym.

A result of 53 or more is very unusual. An exact count in this set of 1000 trials shows that only 30 of them yielded 53 or more ones. This represents a p-value of 0.03, so we conclude that the pilot's improvement in the conversion rate is real and not due to chance.

Formula Procedure

The conventional formula procedure used here by statistical software would be a one-sample Normal approximation to the binomial. In your software program, you would most likely specify a hypothesis test for a single proportion.

The null hypothesis is that the sample proportion is equal to a specified proportion. In the example described earlier, this specified proportion is the gym's usual conversion rate of 25%. The alternative hypothesis is that the conversion rate is greater than 25%.

To interpret your software output, you will need to have a good understanding of these concepts and other ways of expressing them. The actual formula used may lie behind the scenes and is not necessary to obtain the output but is provided here in an appendix.

9.2   A SINGLE MEAN

Returning to the example with humidifier vendors that we studied in a previous chapter, this time we will examine only the sample from Vendor A. This vendor is contractually obligated to deliver humidifiers that average at least 14 oz of moisture output per hour. The following sample shows an average of only 13.84 oz/h. Does this sample prove that Vendor A is out of compliance with its contract? Or might the true average be 14 oz/h and could this low sample average happen by chance? (Table 9.1).

TABLE 9.1   Humidifier Moisture Output (oz/h)

Vendor A
14.2
15.1
13.9
12.8
13.7
14.0
13.5
14.3
14.9
13.1
13.4
13.2
Mean: 13.84

So far, we have been testing hypotheses by creating a box with slips of paper, corresponding to the null model. In this example, it is not immediately clear exactly what would be in that box—it would need to have a mean of 14 oz.

We noted earlier that it is possible to use a confidence interval to test a hypothesis, and this is the simplest way to answer this question. We want to know "does 14 fall within the confidence interval?"

Resampling Procedure for the Confidence Interval

1. Place the moisture output values in a box.
2. Draw out one value at a time with replacement until 12 values have been drawn.
3. Find the mean of the resample.
4. Repeat steps two through three 1000 times.
5. Find the 90% confidence interval and draw a histogram.

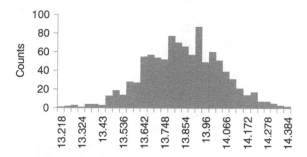

Figure 9.2   Histogram for resampling procedure: 14 is not rare.

The 90% confidence interval for one set of 1000 trials was in the range 13.52–14.17. The histogram for the output is shown in Figure 9.2.

Question 9.1

Reflecting back on the original question (is Vendor A out of compliance?), interpret the results.

Question 9.2

Our goal in this problem is to get a confidence interval—an idea of how much one sample might vary from another. How is it that simply drawing resamples with replacement from the sample itself provides us with useful information?

Formula Approach for the Confidence Interval

The formula approach to this problem, which is used by most statistical software, relies on the fact that a sample mean is distributed according to Student's t distribution, provided the sample is large enough or sufficiently Normally distributed. In StatCrunch, the procedure is to enter the data, then Stat > t-statistics > one sample > with data and then choose the confidence interval and the appropriate level (90%). The results:

L. Limit U. Limit

13.48 14.21

The actual formula is supplied in an appendix.

The resampling and formula methods are roughly equivalent procedures and either may be used.

9.3   MORE THAN TWO CATEGORIES OR SAMPLES

Experiments are expensive to set up. Clinical trials for new drugs can cost hundreds of millions of dollars. Once the design is set, the data are collected and the experiment is complete, which can take years in medical experiments, the results are what they are. It is not possible to go back and revise the design. For this reason, it often makes sense to investigate several things at once.

Count Data

Let us first look at an example with count data.

Example: Marriage Therapy   Do behavioral and insight therapies for marriage counseling differ in effectiveness? Behavioral therapy stresses the skills of managing interpersonal relationships and insight therapy stresses working out underlying difficulties. Fifty-nine couples were randomly assigned, with 29 to the behavioral therapy group and 30 to the insight therapy group. At a 4-year follow-up, 15 of the behavioral group were happily married (HM), three were distressed married (DM), and 11 were divorced (DIV). The insight group had 24 HM, five DM, and one DIV: (Data from Douglas K. Snyder, Robert M. Wills, Arveta Grady-Fletcher, "Long-Term Effectiveness of Behavioral Versus Insight-Oriented Marital Therapy", *Journal of Consulting and Clinical Psychology*, 1991, v. 59, No. 1, 138–141) (Table 9.2).

TABLE 9.2   Marriage Therapy

	Happily Married	Distress Married	Divorced
Behavioral	15	3	11
Insight	24	5	1

Definition: **Contingency table**

In a contingency table, such as the one mentioned earlier, each cell count is contingent on both the row variable—behavioral or insight therapy—and the column variable—state of marriage.

Are the differences among the groups significant?

We could ask whether insight therapy produces more "HM" results but that leaves out the "distress married." Moreover, which is better, "distress married" or "divorced"? Even if we could say which is better, we still do not have a simple comparison of the two groups that can be boiled down to a single statistic.

A single statistic is necessary for a hypothesis test, which in the end will have to answer the question "might this have happened by chance?" where "this" is the extreme value of some statistic.

So, we can step back and ask a more general question: Do the observed results depart from what we would expect to get by chance if the choice of therapy had no effect at all on outcome?

Departure from Expectation

The concept of "departure from expectation" is an important one in statistical inference. By this we really mean "departure from what we would expect if only chance variation were at work."

We have already been testing hypotheses to see if a mean or proportion is bigger (smaller) in one sample than in another or bigger (smaller) than a benchmark. Departure from expectation is a more general overlapping concept that simply asks whether observed sample results (from two or more samples) are *different* from what we would expect in a chance model.

The fundamental idea is to figure out what a sample result (e.g., proportion or mean) would be in a chance model, what it was in actuality, and subtract. The concept can be applied to both continuous (measured) data and count data.

For continuous data, the traditional approach to measuring departure from expectation is *Analysis of Variance* (ANOVA), and this is dealt with in Chapter 13.

For count data, the traditional approach is a *chi-square test*, which we take up next.

The first task in this test of independence is to establish what we would expect if both therapies yielded the same results. We would expect the overall behavioral/insight split to be the same across all marital outcomes. With 29 in the behavioral group and 30 in the insight group, if therapy made no difference, we would expect each outcome (HM, distressed, divorced) to be almost equally split between behavioral and insight therapy (with a bit more to the insight group).

There were 39 HMs. Specifically, we would expect just under half, or $29/59 = 49.2\% = 19.17$ couples, to be in the behavioral group, and just over half of them, or $30/59 = 50.8\% = 19.83$ couples, to be in the insight group. It is the same for DM and divorced.

Alternate option: With equivalent arithmetic, you could note that there were 39 HMs (66.1%), 8 DMs (13.6%), and 12 DIVs (20.3%). So, of the 29 couples in the behavioral group, $66.1\% = 19.17$ couples would be expected to be HM, and so on (Table 9.3).

TABLE 9.3   Expected Outcomes if Treatments Yield the Same Results

	Happily Married	Distress Married	Divorced
Behavioral	19.17	3.94	5.90
Insight	19.83	4.06	6.10

The next step is to determine the extent to which the observed results differ from the expected results. Direction does not matter so we take absolute values.

TABLE 9.4   Absolute Difference Between Observed and Expected Outcomes

	Happily Married	Distress Married	Divorced
Behavioral	4.17	0.94	5.11
Insight	4.17	0.94	5.11

Overall, these differences sum to 20.42, although the Table 9.3 sums to 20.44 due to rounding.

The Key Question

Is this a greater sum of differences than we might expect from a random allocation of outcomes to 29 couples of the behavioral group and 30 couples of the insight group?

To answer this question, we use the following resampling procedure (also called a *permutation procedure*). Resampling Stats uses the term "*urn*" instead of "box."

1. Fill a single urn with 39 ones (HM), 8 twos (distress married), and 12 threes (divorced).
2. Shuffle the urn and take two samples without replacement of sizes 29 and 30.
3. Count the number of ones, twos, and threes in each sample.
4. Reconstruct the resampled counterparts to Tables 9.2 and 9.4 and record the resampled statistic of interest—the sum of absolute differences – observed value = 20.42.
5. Repeat steps two through four 10,000 times.
6. Determine how often the resampled sum of differences exceeds the observed value of 20.42.

The outcome of one resampling trial can be seen in the following section; we shuffle the urn and then treat the first 29 values as the $N = 29$ resample and the remaining 30 as the $N = 30$ resample.

Resampling Stats Output

Shuffled Urn				
3		Resampled table		
1		Ones	Twos	Threes
1	Counts in the first 29:	21	4	4
3	Counts in remaining 30:	18	4	8
2				
1				
1		Expected table		
1		19.17	3.94	5.9
1		19.83	4.06	6.1
1				
2		Absolute differences		
1		1.83	0.06	1.9
1		1.83	0.06	1.9
1				
1		Sum	7.58	
3				
1				
1				
1				
1				
1				

Shuffled Urn
1
2
3
2
1
1
1
1
1
1
1
3
1
2
3
3
1
2
3
1
2
1
1
3
1
3
1
1
1
3
1
1
1
2
1
3
1
1

The statistic of interest here was 7.58, which is not nearly as extreme as the observed value of 20.42. Of course, the above is just one trial. When the simulation was repeated 10,000 times, a statistic ≥ 20.42 was encountered, but only 69 times.

A portion of the relevant 10,000 spreadsheet cells in the vicinity of the observed value of 20.42 is displayed in Figure 9.3 from the sorted output of 10,000 trials.

63	20.66
64	20.66
65	20.66
66	20.66
67	20.42
68	20.42
69	20.42
70	19.58
71	19.58
72	19.58

Figure 9.3   Marriage therapy differences.

We can see that only 69 cells (one through 69) out of all 10,000 trials—fewer than 1%—yielded a sum of differences as big as or bigger than the observed value. This represents an estimated p-value of 0.0069. The chance occurrence of the observed value is so rare that we conclude that there is a real difference among the therapies.

Specific Resampling Stats and StatCrunch software procedures for this example can be found in the textbook supplements.

Question 9.3

Could we resample with replacement instead of without replacement?

Question 9.4

Why use a single urn? Why not separate urns for the behavioral and insight groups?

Chi-Square Test

The statistic used earlier sums the absolute deviations between observed and expected values. Via resampling, this is easy to handle. In the pre-computer age, it was not. Instead, formulas were developed using squared deviations instead of absolute deviations.

Squaring the deviations, rather than taking absolute values, renders them all positive—which we want—while keeping them mathematically tractable for use in formulas.

Definition: **Chi-square statistic**

The chi-square statistic squares each deviation, divides it by the expected value, and then sums the results.

Dividing by the expected value standardizes the distribution so that it is a member of a family of similar chi-square distributions, each characterized by its degrees of freedom. In the days before resampling, this standardization made a comparison possible between the results of a given study like the one mentioned earlier and a tabulated set of chi-square values.

With resampling, the use of absolute values is no problem. Standardizing the statistics is not required because we do not compare the observed value to a tabulated value. Instead, we compare it to values obtained by the resampled distributions.

Chi-Square Example on the Computer Most likely you will be executing similar tests by applying the chi-square procedure using statistical software. An example using StatCrunch is found in the textbook supplements.

Small or Sparse Sample Caution Chi-square software procedures often warn about unreliability in cases where cell frequencies are below certain levels. This is because, as with many statistical procedures, the chi-square formula approximation works well only when sample sizes are large enough. "Large enough" for the chi-square procedure means counts of at least five in each cell.

The resampling test outlined earlier is a general, all-purpose procedure. It is valid for all cell counts. A chi-square procedure is a standardized, math-friendly version that works in most cases and was developed long before the age of computing made resampling procedures easy.

Alternate Terminology We have outlined both a resampling test and a chi-squared formula approximation.

A permutation test is roughly the same thing as the resampling procedure that we have outlined earlier. In its classic form, a permutation test systematically lists all the possible ways that data could be rearranged under the null hypothesis. The procedure that we use for repeatedly and randomly shuffling the data achieves largely the same thing. The more shuffles you do, the closer the shuffling version will get to the classic version.

9.4 CONTINUOUS DATA

The treatment of more than two samples with measured data is handled in the chapter on ANOVA, where it leads into the discussion of multiple linear regression.

9.5 GOODNESS-OF-FIT

In 1991, Tufts University researcher Thereza Imanishi-Kari was accused of fabricating data in her research. Congressman John Dingell became involved, and the case eventually led to the resignation of her colleague, David Baltimore, from the presidency of Rockefeller University.

Imanishi-Kari was ultimately exonerated after a lengthy proceeding. However, one element in the case rested on statistical evidence regarding the expected distribution of digits in data produced by a wide variety of sources.

For example, looking at data from many different natural and man-made sources, it turns out that one occurs as the initial digit in numbers almost twice as frequently as any other digit. When two through nine occur as the initial digit of a number, the frequency of their occurrence steadily declines. See Figure 9.4 and Google Benford's Law for more information.

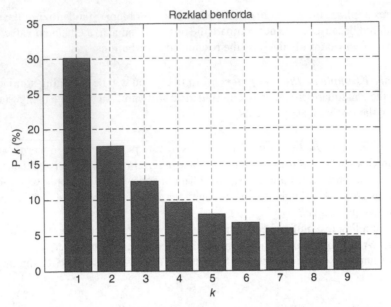

Figure 9.4 The frequency (*y*-axis) of leading digits (*x*-axis) in numbers. Source: Wikipedia article on Benford's Law. The figure is in the public domain.

This is not true of interior digits—they are expected to show a uniform distribution, with each one occurring with a probability of 1/10. Imanishi-Kari's data were examined, and the investigator concluded that the distribution of her interior digits strayed too far from the expected uniform distribution.

TABLE 9.5   Frequencies of 315 Interior Digits

Digit	Frequency
0	14
1	71
2	7
3	65
4	23
5	19
6	12
7	45
8	53
9	6

Source: *Science*, March 8, 1991, News & Comment, p. 1171.

What is too far? Table 9.5 displays the frequencies of 315 interior digits in one of Imanishi-Kari's data tables.

The distribution looks unbalanced, but could the imbalance arise naturally through random chance? Let us test the proposition that the digits 0 through 9 all occur with equal probability. Note how we measure the difference between the resampled distribution and the expected uniform distribution, that is, the expected distribution is that each digit will occur $315/10 = 31.5$ times. In Imanishi-Kari's table, we got 14 zeros when we expected 31.5, so the absolute difference is 17.5. For the ones, the absolute difference is $71 - 31.5$ or 39.5.

Overall, these absolute deviations sum to 216. Is this a greater overall deviation than might be explained by chance?

Resampling Procedure

1. Create an urn with the digits 0, 1, 2, ... , 9

2. Sample with replacement 315 times. The probability of getting a specific digit must remain the same from one draw to the next, hence the need to replace. We need to use Resampling Stats or StatCrunch for this problem because of Box Sampler's sample size limit of 200.

3. Count the number of 0s, 1s, 2s, 3s, and so on. The histogram shown in Figure 9.5 illustrates just one group of 315 digits, using a bin width of one. It looks more balanced than Imanishi-Kari's.

4. Find the absolute difference between the number of 0s and 31.5, the number of 1s and 31.5, and so on, and sum.

5. Repeat steps 2–4, 10,000 times.

6. How often did the resampled sum of absolute deviations equal or exceed 216? Divide this sum by 10,000 to calculate the p-value.

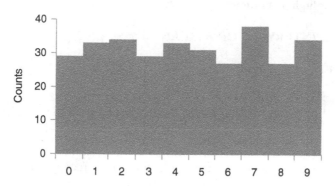

Figure 9.5   Histogram of one 315-digit resample.

Results: Not once in 10,000 trials did an imbalance as great as the observed imbalance occur, for an estimated p-value $= 0.0000$.

These results show that an imbalance as great as the observed one is *extremely* rare, so we reject the null hypothesis that chance is responsible. This does not prove that Imanishi-Kari invented the results; it is possible that some other nonchance mechanism could be at work.

This statistical procedure is known as a "goodness-of-fit" test. It examines how well an observed distribution fits a theoretical expectation.

APPENDIX: NORMAL APPROXIMATION; HYPOTHESIS TEST OF A SINGLE PROPORTION

Let us denote the sample size as n, the proportion in the sample as p, and the specified proportion as P_0.

The sample statistic is given by

$$Z = \frac{p - P_0}{\sqrt{\frac{P_0(1-P_0)}{n}}}.$$

Z follows the standard Normal distribution.

If we use percentages instead of proportions, the sample statistic described earlier becomes

$$Z = \frac{p - P_0}{\sqrt{\frac{P_0(100-P_0)}{n}}}.$$

$-Z_{\alpha/2}$ and $Z_{\alpha/2}$ are the $\alpha/2$ percentile and the $(1 - \alpha/2)$ percentile, respectively.

Using the sample data, we calculate the value of the sample statistic Z.

If calculated $Z < -Z_{\alpha/2}$ or $Z > Z_{\alpha/2}$, we reject the null hypothesis and conclude at significance level α that the proportion in the sample is not equal to the specified proportion.

If $-Z_{\alpha/2} < Z < Z_{\alpha/2}$, then we do not reject the null hypothesis and we conclude that the proportion in the sample may be equal to the specified proportion.

The Normal approximation to the binomial does not work with small samples and/or extremely low or high proportions.

CONFIDENCE INTERVAL FOR A MEAN

The $100(1 - \alpha)\%$ t-interval for a single mean is

$$= \left(\overline{X} - t_{n-1,\ \alpha/2} * \frac{s}{\sqrt{n}}, \overline{X} + t_{n-1,\alpha/2} * \frac{s}{\sqrt{n}} \right),$$

where n is the sample size, $\overline{X}$ is the sample mean, and $-t_{n-1,\ \alpha/2}$ and $t_{n-1,\ \alpha/2}$ are the t-values with $n - 1$ degrees of freedom corresponding to the $(\alpha)/2$ percentile and the $(1 - \alpha/2)$ percentile.

For the humidifier example presented earlier $n = 12$, $\overline{X} = 13.84$, $s = 0.704$, and the fifth and 95th percentiles are -1.796 and 1.796 using $\alpha = 0.10$.

Hence, the 90% t-interval for the mean is

$$= \left(13.84 - 1.796 * \frac{0.704}{\sqrt{12}},\ 13.84 + 1.796 * \frac{0.704}{\sqrt{12}} \right)$$

$$= (13.84 - 0.365,\ 13.84 + 0.365)$$

$$= (13.475,\ 14.205)$$

9.6 EXERCISES

1. A nursery stock breeder breeds flowers and believes, based on reading a book on genetics, that a certain cross will produce plants with either white, red, or pink flowers, in the following proportions: 60% white, 30% red, and 10% pink. For each plant, the flowers are always of one color or another. The breeder is experimenting with a treatment that may alter the above proportions. In a test of 100 plants, 65 produced white flowers, 24 produced red flowers, and 11 produced pink.
   State and test an appropriate hypothesis. You may use software of your choice.

2. A social media site that allows community members to rate restaurants wants to ensure that the reviews are genuine reviews of the community and not favorable reviews orchestrated by the friends of the restaurant owners. It uses several metrics when examining reviews, one of which is the "percent favorable." It theorizes that a spike in the percent favorable might represent a campaign by the restaurant, but also recognizes that it might reflect genuine customer sentiment, or just random variation. So it collects 4 days' worth of such data on a periodic basis and subjects it to a hypothesis test. For one restaurant, the percent favorable has stood at 60% for the last year. One recent 4-day sample showed 72% favorable—23 favorable reviews and 9 unfavorable.

   (a) Specify and conduct an appropriate hypothesis test.

   (b) In a couple of sentences, interpret the results of your hypothesis test.

   (c) Given the three possible causes of a spike in favorable ratings (restaurant campaign, actual customer sentiment, and random variation) and discuss the role of a hypothesis test in distinguishing among the three causes.

3. A software service firm has been growing relatively rapidly and concentrating on winning contracts and delivering on them. It has paid less attention to getting paid on time, and the investors who own the firm have established a target of $2500, maximum, on the average level of accounts receivable (the average per account, not the total across all accounts). If it is clear that this average is being exceeded, they want a corrective action. On the other hand, they agree with the management that the primary focus should be on winning contracts and delivering on them. So it has been decided to take periodic samples of accounts receivable to watch for clear evidence that the target is being exceeded. Here is the most recent one (in $):
   950, 3200, 1230, 1500, 5500, 4500, 6340, 2900, 5375, 4300, 2200, 3100

   (a) Specify and conduct an appropriate hypothesis test.

   (b) In a couple of sentences, interpret the results of your hypothesis test.

   (c) Could the simple point estimate from the sample be used, without doing a hypothesis test? If that were done, what would be the effect on "false positives?"

4. Consider the following data on two therapies and their outcome frequencies in a study:

	No Improvement	Some Improvement	Cure
Therapy A	7	5	2
Therapy B	2	6	6

(a) Which therapy seems to do better?

(b) Specify and conduct an appropriate hypothesis test.

(c) In a couple of sentences, interpret the results of your hypothesis test.

Answers to Questions

9.1 The average moisture output that the vendor has promised to supply is 14, and this lies within the 90% confidence interval. Therefore, the low sample average could have happened by chance.

9.2 (Our goal is to get a confidence interval ...):

If the entire population of moisture output values were available to us, we could draw actual samples to see how much they vary from one another. But we do not (and, if we did, we would not need to take a sample anyhow). The next best thing is to create a hypothetical population and draw samples from it, to see how they behave.

One way to do this is to replicate the observed sample, say, thousands or millions of times, and then draw samples from it. This has the virtue of capturing all the information we have about the population—the information in the sample itself, without adding additional ingredients.

A shortcut is to simply draw the resamples with replacement—that is the same thing as replicating the sample an infinite number of times and drawing without replacement. The probability of drawing, for example, the 14.2 is 1/12 on each draw, in either case.

9.3 Yes. The two procedures have slightly different statistical properties that are beyond the scope of this text but either could be used.

9.4 (why use a single urn ...):

We want to test the null model that the two groups share the same proportion of HMs, DMs, and DIVs and that differences between insight and behavioral are the result of chance. We do this by having a single urn that has all the HMs, DMs, and DIVs together. Separate urns with the different HMs, DMs, and DIVs for the two groups would not be testing the proposition that they share common rates.

10

CORRELATION

Think back to the "no fault" study of errors in hospitals. In an experiment, we found that introducing a no-fault reporting system reduced the number of serious errors. We found that there was a relationship between one variable—the type of reporting system—and another variable—a reduction in errors.

The "type of reporting system" is a binary variable. It has just two values: "regular" and "no-fault."

Often, you may want to determine whether there is a relationship involving the *amount* of something, not just whether it is "on" or "off."

For example, is there a relationship between employee training and productivity? Training is expensive, and organizations need to know not simply *whether* training helps but also *how much* it helps.

After completing this chapter, you will be able to:

- explain how the vector product sum measures correlation,
- use a resampled vector product sum to test the statistical significance of a measured correlation,
- explain the useful properties of the correlation coefficient, and how it is a standardized version of the vector product sum,
- perform a resampling test of the correlation coefficient,
- state some limitations and cautions when measuring correlation.

Definition: Correlation

Correlation is an association between the magnitude of one variable and that of another—for example, as x_1 increases, x_2 also increases. Or as x_1 increases, x_2 decreases.

Introductory Statistics and Analytics: A Resampling Perspective, First Edition. Peter C. Bruce.
© 2015 John Wiley & Sons, Inc. Published 2015 by John Wiley & Sons, Inc.

Different statistics are used to measure correlation; we will develop and define two in this chapter.

10.1   EXAMPLE: DELTA WIRE

In the 1990s, Delta Wire, a small manufacturing company in Clarksdale, Mississippi, started a basic skills training program. Over time, as wastage declined, production at the facility improved from 70,000 pounds of wire per week to 90,000 pounds per week. Table 10.1 shows some detailed data from the program.

TABLE 10.1   Training and Productivity at Delta Wire

Total Training Hours, Cumulative	Productivity, Pounds Per Week
0	70,000
100	70,350
250	70,500
375	72,600
525	74,000
750	76,500
875	77,000
1100	77,400
1300	77,900
1450	77,200
1660	78,900
1900	81,000
2300	82,500
2600	84,000
2850	86,500
3150	87,000
3500	88,600
4000	90,000

(The Delta Wire case was reported in Bergman, Terri, "TRAINING: The Case for Increased Investment," *Employment Relations Today*, Winter 1994/95, pp. 381–391, and is available at http://www.ed.psu.edu/nwac/document/train/invest.html. Detailed data are from an adaptation by Ken Black, *Business Statistics*, Wiley, Hoboken, NJ, 2008, p. 589.)

Clearly, more training led to higher productivity throughout the period. We can get a better idea of the relationship from the scatterplot in Figure 10.1. Henceforth, we will deal with productivity in terms of thousands of pounds per week for easier display.

Figure 10.1 presents a clear picture: more training is associated with higher productivity. No further analysis is needed to reach this conclusion. We can say that the correlation is positive, and the relationship is more or less linear.

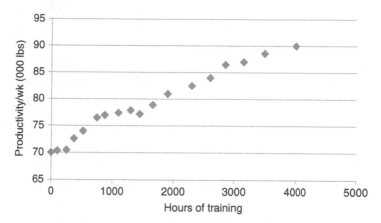

Figure 10.1   Delta Wire productivity versus training.

10.2   EXAMPLE: COTTON DUST AND LUNG DISEASE

Figure 10.2, which illustrates the respiratory function of workers in cotton factories in India, shows a less clear picture. The *y*-axis shows a worker's Peak Expiration Flow Rate (PFER) and higher is better. The *x*-axis shows the years of exposure to cotton dust in a cotton factory.

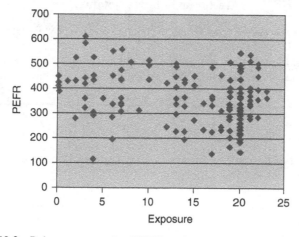

Figure 10.2   Pulmonary capacity (PEFR) and exposure to cotton dust (years).

(The diagram is from "To Know the Prevalence of Byssinosis in Cotton Mill Workers & to know Changes in Lung Function in Patients of Byssinosis," *Indian Journal of Physiotherapy and Occupational Therapy,* Authors: Sarang Bobhate, Rakhi Darne, Rupali Bodhankar, Shilpa Hatewar, Vol. 1, No. 4 (2007-10–2007-12); available as of July 2011 at http://www.indmedica.com/journals.php?journalid=10&issueid=103&articleid =1424&action=article)

Is PEFR related to Exposure? It is hard to say just based on the picture. We need the following:

1. A way to measure correlation between two numeric variables, and
2. A way to determine whether the correlation we measure is real or might just be a chance association.

We will tackle these questions in Section 10.3. But before we proceed:

Question 10.1

Can you spot an unusual feature of the exposure variable? What might account for it?

10.3   THE VECTOR PRODUCT AND SUM TEST

Vectors **x** and **y** are correlated if high values of **x** correspond to high values of **y**. One way to measure correlation between **x** and **y** is to multiply them together as vectors and then to sum them up. This sum is the *vector product sum,* and it is greatest when the rank order of **x** matches that of **y**.

Vector and Matrix Notation

A *vector* is a list of numbers—a single column or row. It is denoted by a lower-case bold letter, as with **x** and **y** shown earlier.

A *matrix* is a two-dimensional array of numbers with both rows and columns. It is denoted by an upper-case bold letter, for example, **X**. We have not yet dealt with matrices.

Definition: **Rank order**

Rank order is a list of the rank positions of a list of values. For example, consider the list {7 9 6 3}. The rank order of this list is {3 4 2 1}.

Try It Yourself 10.1

Let **x** be [1 2 3] and **y** be [2 3 4]. Note that they are in the same rank order—lowest to highest. When multiplied together, the vector product sum is 20.

x	y	Product
1	2	2
2	3	6
3	4	12

Vector product sum $= 2 + 6 + 12 = 20$
Let's try a different arrangement:

x	y	Product
2	2	4
1	3	3
3	4	12

Vector product sum $= 4 + 3 + 12 = 19$

Try rearranging the **x** vector in various other ways and then recalculating the vector product sum. You will see that it is never as high as it is when the two vectors are in the same rank order. Note: In the same way, the vector product sum is smallest when the rank orders of the two variables are exact opposites—that is, when they are perfectly negatively correlated.

Definition: **Perfect correlation**

Perfect positive correlation is when the rank order of one variable exactly matches the rank order of the second variable—highest value goes with the highest value, second-highest with the second-highest, and so on. Perfect negative correlation is when the rank order of one variable is the exact opposite of the second variable—the highest value goes with the lowest, the second-highest goes with the second-lowest, and so on.

Your "Try it Yourself" results with the arrangements of [1 2 3] and [2 3 4] suggest a way of determining whether an apparent correlation between two variables might have happened by chance.

1. Write down the values for one of the variables on a set of cards.
2. Write down the values for the other variable on a second set of cards.
3. Array the two sets of cards next to each other; one column for variable one and the other column for variable two. Make sure that cards for the same case are adjacent to one another. For example, in the lung disease case, the card for 110 PERF score must be next to a card for 4-year exposure. Multiply together these two variables and sum these values to calculate the vector product sum.
4. Shuffle one set of cards, repeat the multiplication, and record the vector product sum.
5. Repeat step two 1000 times.
6. Find out how often the shuffled sum is equal to or greater than the observed sum.

The result is the p-value. Given the null model of no correlation, the p-value is the probability that a vector product sum as large as or larger than the observed value might occur by chance.

Example: Baseball Payroll

Is the total salary payroll of a team correlated with performance? More specifically, do teams with higher payrolls win more games? (Figure 10.3).

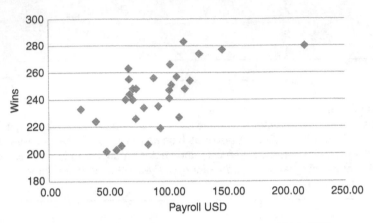

Figure 10.3   Baseball payroll versus total wins, 2006–2008.

Table 10.2 shows the actual data for several teams, along with the products and the vector product sum.

TABLE 10.2   Excerpt of Baseball Payroll and Total Wins

Team Name	Average Payroll (Million $)	Total Wins	Product
Yankees	216.1	279	60,292
Red Sox	146.66	276	40,478
Mets	127.4	273	34,780
Rays	39.76	224	8906
Marlins	27.07	233	6307
			668,620

Resampling Procedure

1. Multiply the payroll vector by the wins vector. The sum of products is 668,620.
2. Shuffle one vector—we will shuffle the wins vector—and then re-multiply. Record the shuffled sum.
3. Repeat step 2 many times (say 1000).
4. How often do we get a sum of products ≥668,620?

From the histogram shown in Figure 10.4, we can see a product sum of 668,620 rarely, if ever occurs. A detailed examination of the results in the spreadsheet file shows that it did not happen once in a set of 1000 simulations. So we conclude that the degree of correlation between payroll and wins is not just coincidence.

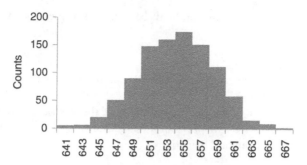

Figure 10.4   Baseball histogram of shuffled product sums (000).

There are Resampling Stats and StatCrunch procedures for the baseball payroll problem in the textbook supplements.

Data file download: baseball_payroll.xlsx.

The following *Try It Yourself* can be found in the textbook supplements.

Try It Yourself 10.2

(Optional—requires Resampling Stats or StatCrunch)

An online dating site seeks to learn more about what it can do to encourage successful outcomes to the relationships formed by its customers. It collects various data, including satisfaction surveys 6 months after a customer first signs up and also how much time the customer has spent on its dating site. Satisfaction data is recorded as an integer between 1 and 10, and time-on-site is recorded in minutes. Here are the hypothetical results:

Time spent	Satisfaction
10.1	2
67.3	7
34	2
2.9	1
126.3	9
39	8
4.6	1
211.3	6

Calculate the vector product sum and use a resampling procedure to test whether there is a correlation between time spent and satisfaction.

10.4   CORRELATION COEFFICIENT

The vector product sum can be used to test the statistical significance of correlation between two variables, but the sum itself is not that meaningful.

Consider the vector product sum for the three examples as mentioned earlier.
Baseball: 688,620.
Lung disease: 603,670.
Worker training: 2,397,534.
These numbers can neither be compared to one another nor can they be used to measure the strength of the correlation.

Instead, statisticians use a standardized version of the vector product sum called the *correlation coefficient*. As we saw previously, standardization involves subtraction of the mean and division by the standard deviations.

Definition: Correlation coefficient

Let X and Y denote the two variables. Consider a sample of size n of paired data (X_i, Y_i). Then, the sample correlation coefficient, denoted as r or the Greek letter ρ (rho), is given by

$$r = \frac{\sum_{i=1}^{n}\left(X_i - \overline{Y}\right)\left(Y_i - \overline{Y}\right)}{(n-1)s_x s_y},$$

where $(\overline{X}, \overline{Y})$ are the sample means and (s_x, s_y) are the sample standard deviations for X and Y, respectively. The correlation coefficient formula can also be given by:

$$r = \frac{\sum_{i=1}^{n}\left(X_i - \overline{X}\right)\left(Y_i - \overline{Y}\right)}{\sqrt{\sum_{i=1}^{n}\left(X_i - \overline{X}\right)^2 \sum_{i=1}^{n}\left(Y_i - \overline{Y}\right)^2}}.$$

As a result of standardization, the correlation coefficient always falls between -1, which is a perfect negative correlation, and $+1$, which is a perfect positive correlation. A value of zero indicates no correlation.

Note: The correlation coefficient defined above is called the *"Pearson product moment correlation coefficient"* or simply the *"Pearson correlation coefficient."* It is named after the great English statistician, Karl Pearson (1857–1936). There are other correlation coefficients, but this is the most widely used.

The correlation coefficients for the three examples cited earlier are as follows. The x–y scatterplots are repeated for reference (Figure 10.5).

1. Baseball: $\rho = 0.64$ (Figure 10.5)
2. Lung disease: $\rho = -0.28$ (Figure 10.6)
3. Worker training: $\rho = 0.99$ (Figure 10.7)

For the worker training example, the 0.99 value for the correlation coefficient indicates a very strong correlation between training time and productivity.

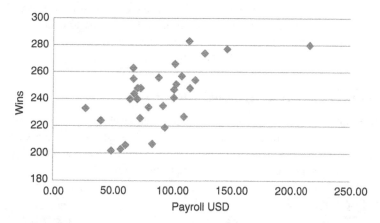

Figure 10.5   Baseball payroll versus total wins, 2006–2008 $\rho = 0.64$.

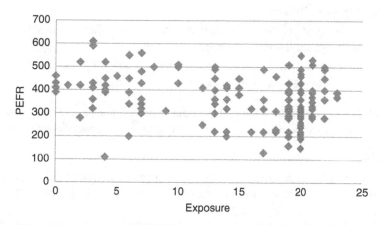

Figure 10.6   Pulmonary capacity (PEFR) and exposure to cotton dust (years) $\rho = -0.28$.

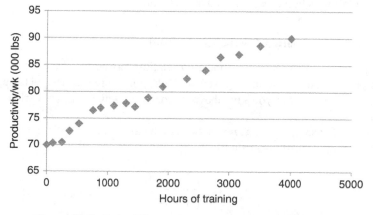

Figure 10.7   Delta Wire productivity versus training $\rho = 0.99$.

For the lung disease example, the -0.28 value for the correlation coefficient indicates a modest negative correlation between exposure time and lung function.

For the baseball example, the 0.64 value for the correlation coefficient indicates a strong correlation between payroll and won-loss record.

Inference for the Correlation Coefficient—Resampling

Could the correlation simply be the result of random chance?

Significance testing for correlation has already been illustrated for the vector product sum method. We can use the same procedure for hypothesis tests except that we calculate the correlation coefficient instead of the vector product sum.

Hypothesis Test—Resampling

1. Calculate the correlation coefficient for the two variables.
2. Shuffle one of the variables, calculate this shuffled correlation coefficient, and record.
3. Repeat step two 1000 times.
4. When the observed correlation is positive, find out how often the shuffled correlation coefficient is greater than the observed value. OR ...
5. When the observed correlation is negative, find out how often the shuffled correlation coefficient is less than the observed value.

The result is the *p*-value. Given the null model of no correlation, the *p*-value is the probability that a correlation coefficient as least as extreme as the observed value might occur by chance.

Example: Baseball Using Resampling Stats   The specific Resampling Stats procedure for the baseball example is in the textbook supplements. Figure 10.8 illustrates the results of one experiment.

Under the null hypothesis of no correlation, the resampling (shuffled) distribution of the correlation coefficient for these data centers, as expected, around 0. It rarely exceeds 0.5, so we can be safe in concluding that the value of 0.64 did not arise by chance and that the apparent correlation between wins and payroll is real.

Inference for the Correlation Coefficient: Formulas

The null hypothesis is that there is no linear correlation between X and Y in the population. Before computer-intensive resampling was available, a formula was needed to approximate the shuffled distribution that you saw above. Student's t, which we saw earlier, is used in this formula.

Let us denote the sample size as n and the sample correlation coefficient as r.

Then, the sample statistic is given by

$$t = \frac{r\sqrt{n-2}}{\sqrt{1-r^2}}.$$

t follows Student's t distribution with $n-2$ d.f. (degrees of freedom).

Menu Commands | Toolbar Commands | Custom Toolbars

B1 =PERCENTILE(A1:A1000,0.05)

	A	B	C	D	E	F
1	0.075289	-0.30908	<- 5% Level			
2	0.205848	0.301923	<- 95% Level			
3	0.122352					
4	-0.00392					
5	0.11708					
6	0.260676					
7	0.164731		Bin MidPt	Counts	% Total	Cu. Freq.
8	0.111016		-0.454	8	0.8	0.8
9	0.121419		-0.383	23	2.3	3.1
10	0.197204		-0.313	47	4.7	7.8
11	-0.37112		-0.242	58	5.8	13.6
12	-0.10757		-0.171	93	9.3	22.9
13	-0.04334		-0.101	128	12.8	35.7
14	-0.24146		-0.03	156	15.6	51.3
15	0.14517		0.041	146	14.6	65.9
16	0.001124		0.112	123	12.3	78.2
17	-0.02405		0.182	103	10.3	88.5
18	-0.05053		0.253	53	5.3	93.8
19	-0.1523		0.324	37	3.7	97.5
20	-0.11142		0.394	14	1.4	98.9
21	0.190935		0.465	8	0.8	99.7
22	0.067068		0.536	2	0.2	99.9
23	0.012833		0.607	1	0.1	100
24	0.093737					

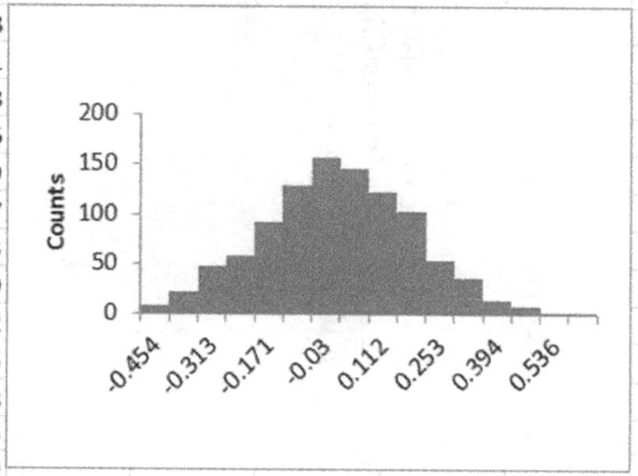

Figure 10.8 Resampling distribution of correlation coefficient for baseball under the null hypothesis.

$-t_{n-2,\alpha/2}$ and $t_{n-2,\alpha/2}$ are the t-values corresponding to the $\alpha/2$ percentile and the $(1 - \alpha/2)$ percentile, respectively.

Using the sample data, we calculate the value of the sample statistic t.

If the calculated value of $t < -t_{n-2,\alpha/2}$ or $t > t_{n-2,\alpha/2}$, we reject the null hypothesis and we conclude at significance level α that the population does have some correlation.

If $-t_{n-2,\alpha/2} < t < t_{n-2,\alpha/2}$, then we do not reject the null hypothesis, and we conclude that there may not be any correlation in the population.

10.5   OTHER FORMS OF ASSOCIATION

So far, we have been considering a relationship between **x** and **y** in which high values of **x** correspond to high values of **y**, or vice versa, on a consistent basis.

Now, consider these hypothetical data on the relationship between tax rates and tax revenue in the United States (Figure 10.9).

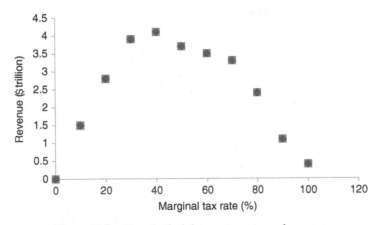

Figure 10.9   Hypothetical data on tax rates and revenue.

At a zero tax rate, of course, no revenue is collected. As the tax rate, shown on the x-axis, increases, tax revenue, shown on the y-axis, also increases. However, at a certain point, the high tax rates make it increasingly profitable to employ tax avoidance schemes. Tax evasion increases; as a result, the revenue collected flattens out and begins to drop. At the point where the government seeks to take most of your money, the legal economy withers away. Most activity moves to the underground economy to escape taxes, and the tax revenues collected decline precipitously. The earlier described data are hypothetical, but they are based in part on research concerning the "inflection point" where increasing rates do not yield additional revenue. There is much disagreement, arising from differing political perspectives, about the location of this point.

Certainly there is a strong relationship between tax rates and revenue, but it will not show up clearly in the correlation coefficient because low revenue can be associated with both low tax rates and high tax rates. A nonlinear model is required to describe and investigate this relationship, and this is beyond the scope of this text. This case is introduced to illustrate that the absence of a linear correlation does not necessarily demonstrate the absence of an association.

10.6   CORRELATION IS NOT CAUSATION

Correlation—even statistically significant correlation—does not imply anything about causation. Below, we present some examples of cases in which two variables are correlated but causation is nonexistent.

A Lurking External Cause

In 1999, a University of Pennsylvania Medical Center study found that infants who slept with lights on were more likely to develop myopia (near-sightedness) later in life. However, the real cause is not the light but a genetic link to myopic parents. Lights in infants' rooms were more likely to be left on by myopic parents than nonmyopic parents. This was the conclusion of a later study. The first study was published in the May 13, 1999 issue of *Nature*; see http://en.wikipedia.org/wiki/Correlation_does_not_imply_causation.

In this case, the event "lights left on" was correlated with the development of myopia but this was not the cause. The real cause was the parents' myopia. This not only led to the child's myopia but it was also the reason the lights were left on. Myopic parents needed well-lit rooms so they could safely make their way to the infant's crib.

When A causes both B and C, then B and C are correlated.

Coincidence

Consider the rates of murder in the 50 US states and District of Columbia (Figure 10.10).

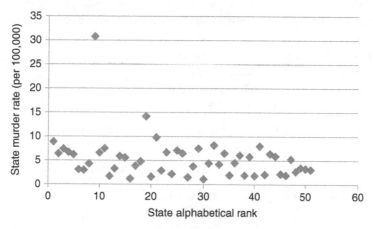

Figure 10.10   Murder rates and alphabetical order of states, $\rho = -0.28$.

The correlation between a state's murder rate and its position in the alphabetical order of states—Alabama being #1 and Wyoming being #50—is modestly negative at −0.28. It is also statistically significant. Yet, there seems no plausible reason why the two should be related, so we are inclined to think that this is due to coincidence.

Statistical significance is not a 100% guarantee that a relationship or finding is real rather than just coincidence. Significance testing reduces the chances of being fooled but it does not eliminate them. The more you examine a given data set in hopes of finding something interesting, the greater is your chance of finding an interesting something that is actually solely due to chance.

Try It Yourself 10.3

Consider the following studies and assess whether the reported correlation is probably part of a cause and effect relationship. Consider whether there is a reasonable theory to explain the correlation and whether some third factor might cause the correlation.

1. A web merchant finds a correlation between time spent on the web site and money spent at checkout.
2. A medical study finds that elderly subjects who walk the fastest live the longest.
3. There is a positive correlation between income and education.
4. An exhaustive review of health records shows that higher consumption of zinc supplements is positively correlated to the scope of a person's social network.

10.7   EXERCISES

1. In the following situations, what form of association would you expect between the variables—linear, nonlinear, or none at all? If linear, positive or negative? And would you expect there to be a causal relationship?
   a. Income and obesity in the United States
   b. Advertising and market share
   c. Cement production and potato production
   d. Web site traffic and web site sales

2. Consider the following two sets of numbers: [1 2 3] and [6 5 4]
   (a) Calculate the vector product sum.
   (b) Repeat the following procedure and calculation five times:

      – Randomly rearrange the values [6 5 4]
      – Recalculate the vector product sum

      Did you ever get a result as small as you did in part (a)?

3. The following questions use this table, which can be downloaded here in Excel format in girls.xls. These data are the heights (in cm) of girls of different ages (in years). For the following questions, you may use software of your choice. (Our source is Siegel and Morgan, *Statistics and Data Analysis: An Introduction*, 2nd. ed., John Wiley and Sons, 1996. Their source is the 1980 World Almanac)

Height	Age
86.5	2
95.5	3
103.0	4
109.8	5
116.4	6
122.4	7
128.2	8
133.8	9
139.6	10
145.0	11

(a) What would be more reasonable here, to think of height as dependent on age or age dependent on height?

(b) Make a scatterplot of these data and describe what you see.

(c) Find the correlation between these two variables (age and height).

(d) It only makes sense to compute a correlation under certain circumstances. Why is it reasonable here?

(e) Interpret the correlation coefficient in terms of the growth of girls.

4. Activity in the portions of the brain connected with social perception was tracked for a sample of university students, and the number of Facebook friends was recorded for the same sample. (From R. Kanai, B. Bahrami, R. Roylance, G. Rees, "Online social network size is reflected in human brain structure," *Proceedings of the Royal Society B*, Published online http://rspb.royalsocietypublishing.org/content/early/2011/10/12/rspb.2011.1959.full)

Review the data in Brain-Facebook.xls, then

(a) Plot the two variables in a scatterplot.

(b) The metric used for GM density is complex to some extent and it is not necessary to understand it for the purposes of this exercise. However, looking at the scale for GM density, make a guess at how those numbers were calculated. Hint: Review Section 4.4.

(c) Calculate the correlation between brain activity and Facebook friends.

(d) Test whether it is statistically significant and interpret the results.

(e) What is the hypothesized direction of causality? Is this direction confirmed by the results of the hypothesis test?

5. In the 1950s, the incidence of polio was on the rise. Disease rates were found to be correlated with the consumption of ice cream (more ice cream, more polio), and some medical authorities advised parents not to feed ice cream to their children. Can you think of an external factor that might be correlated with both ice cream consumption and polio incidence?

Answers to Try It Yourself

10.3

1. This correlation is probably a real cause and effect relationship. It makes perfect sense that spending more money online would require spending more time browsing for products.

2. This might or might not be cause and effect. Many studies show a positive effect of exercise, and exercise does help keep the body in better condition. On the other hand, it could be that a third factor, lack of illness in some elderly, allows faster walking and leads to longer life spans.

3. This correlation is probably a real cause and effect relationship. On balance, education improves skills, and employers are willing to pay more for skilled workers. There is also a possibility of a reverse cause and effect "feedback" relationship—education is expensive, and both families and regions with higher income generate more resources for students to get more and better education.

4. This could be coincidence. There is no apparent theory that explains why there should be a cause and effect relationship between the two factors or any linked third factor. Part of the clue here is that there was an "exhaustive review." It is likely that many possible associations were examined, and only this one, with a significant correlation, was reported. This is like flipping a coin 10 times, repeating the 10 flips over and over, and then just reporting the one set of coin flips that happened to look interesting.

Answers to Questions

10.1 There are a lot of cases clustered at exactly 20 years, compared to other years. Perhaps, workers who had been working for about that long could not remember precisely when they started, and "20 years" was a convenient, round number estimate.

11

REGRESSION

With correlation, all we can measure is the relative strength of an association and whether it is statistically significant. With regression, we can model that association in a linear form and predict values of Y given the values of X.

After completing this chapter, you will be able to

- specify the equation format for a simple linear regression model,
- define residuals (errors),
- fit a linear regression line by eye,
- describe how fitting the regression line by minimizing residuals works,
- use the fitted regression model to make predictions of y, based on the values of x,
- interpret residual plots,
- determine the confidence interval for the slope of a regression line.

The simple form of a linear regression model is as follows:

$$y = ax + b$$

We read this as "y equals a times x, plus a constant b." You will note that this is the equation for a line with slope a and intercept b. The value a is also termed the *coefficient for x* (Figure 11.1). The constant b is where the regression line intersects the y-axis and is also called the y-intercept.

Introductory Statistics and Analytics: A Resampling Perspective, First Edition. Peter C. Bruce.
© 2015 John Wiley & Sons, Inc. Published 2015 by John Wiley & Sons, Inc.

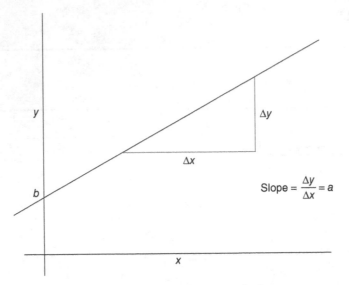

Figure 11.1 Slope and intercept of a line.

11.1 FINDING THE REGRESSION LINE BY EYE

Using the baseball payroll example and assuming that a correlation exists between the payroll amount in dollars and the number of wins over three seasons, can we predict wins based on a given payroll amount?

On the basis of Figure 11.2, it appears that an increase in payroll generally predicts an increase in wins. Suppose we ask the question, "How many wins can be expected over a 3-year period with a payroll of $130 million dollars?" It might be possible to arrive at a reasonable answer based on the scatterplot.

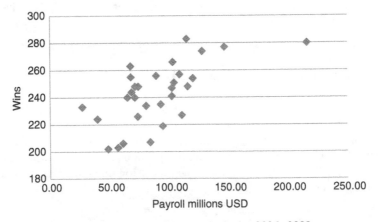

Figure 11.2 Payroll versus total wins 2006–2008.

A better, first step might be to find a line that best represents the data. This is sometimes called a "*line of best fit*" or a "*trend line*." An eyeball estimate of such a line is shown in Figure 11.3. Gridlines have been added to the graph to aid estimation.

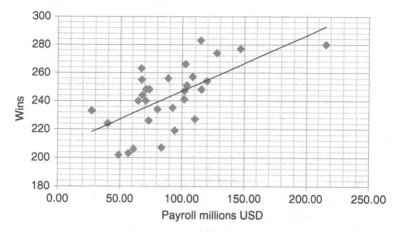

Figure 11.3   Estimated trend line, drawn by eye.

Assuming that the trend line is reasonable, one can estimate approximately 258 wins over three seasons with a payroll of $130 million. The word "approximately" is important. Other factors—including, of course, chance—affect performance, so relatively few teams fall *exactly* on the trend line. The difference between an actual *y* value and the predicted *y* value is called a *residual* or *error*.

Definition: **Residual**

Residuals are the differences between the actual *y* values and the values predicted by the trend line or linear regression equation. In a world of perfect prediction, all residuals would be zero, indicating that no error exists between the prediction and the true value. In reality, this never happens, and there is always some error. The key to finding the optimum trend or regression line is to minimize the error in the residuals.

Residual = $Y - \hat{Y}$ where Y is the actual value and $\hat{Y}$, pronounced "y-hat," is the predicted value.

Definition: **Error**

In statistics, the term "*error*" does not mean "mistake." Rather, it simply means the difference between predicted and actual values. It is the same thing as the residual. An error in a positive direction is just as bad as an error in a negative direction, and to keep positive and negative errors from canceling each other out we use either absolute error or, more commonly, squared error.

Now, we can add an error term to the regression equation. The equation for the regression line is (again):

$$y = ax + b$$

The general equation for predicting y from x is then

$$y = ax + b + e$$

We read this as "y equals a times x, plus a constant b, plus an error e." The constant is the predicted value of y when $x = 0$.

Making Predictions Based on the Regression Line

Let us go one step further. If an equation for the trend line can be found, the estimation of wins based on payroll becomes easier. On the basis of the graph, the trend line goes through or very close to the points (30.00, 220) and (60.00, 232). The slope of this line is 0.40, which means that in general, as payroll increases, so do the number of wins. For this example, the equation for the estimated trend line is

$$y = 0.40x + 208$$

The variable x is the predictor. In this case, x is the payroll in million USD. The y variable represents the predicted number of wins. The constant 208—the y-intercept—is the value the equation takes when x represents zero payroll.

Using \$130 million as the predictor, $0.40(130) + 208 = 260$ wins, which is very close to our eyeball estimate. Predicting values within the limits of the data is called *interpolation*. Predicting values outside the limits of the data is called *extrapolation*.

Question 11.1

What conclusions could you draw from extrapolation in this case?

Placing a trend line by eye based on an esthetic best fit may work in some cases, but a more scientific method is needed to cover all situations.

11.2   FINDING THE REGRESSION LINE BY MINIMIZING RESIDUALS

You will usually use statistical software to calculate the trend line, rather than drawing it by eye. The software does this by finding the line that minimizes the error.

In other words, the computer will find the equation of the line that minimizes the sum of the error terms—the residuals. Figure 11.4 shows the residuals as vertical lines between the regression line and the observations.

In practice, the mathematics of linear regression does not minimize the absolute residual error. Instead, for mathematical convenience, it minimizes the squared residual error. This procedure is called *Least Squares Regression*.

Definition: **Least squares regression**

The least squares regression line for x–y data is the line that minimizes the sum of the squared residuals between the actual y-values and the y values that are predicted by that line.

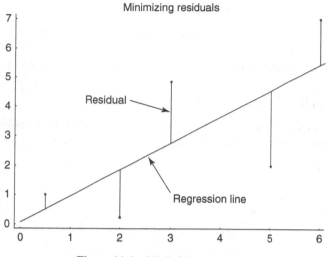

Figure 11.4   Minimizing residuals.

The least squares regression line for the baseball payroll data shown earlier is:

$$y = 0.39047x + 207.4793$$

11.3   LINEAR RELATIONSHIPS

Linear relationships are powerful. If a linear relationship exists between two variables, then it becomes possible to predict the unknown value of one variable based on the known value of another.

Some linear relationships are obvious. Figure 11.5 illustrates the Delta Wire example from the correlation section, with a trend line added. As the number of hours of training increases, so does the productivity. The linear relationship appears to be very strong, and the trend line almost draws itself.

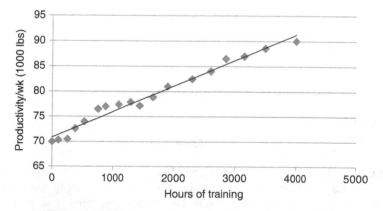

Figure 11.5   Delta Wire hours of training and productivity.

The least squares regression line equation is:

$$y = 5.093445x + 70,880.25$$

When an obvious linear relationship does not appear to exist, what can be done?

Example: Workplace Exposure and PEFR

How and where would a trend line be placed to represent the data shown in Figure 11.6? No linear relationship is obvious. The trend line calculated by Excel is shown in Figure 11.7.

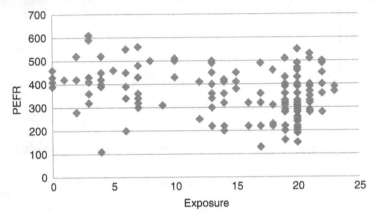

Figure 11.6   Pulmonary capacity (PEFR) and exposure to cotton dust (years).

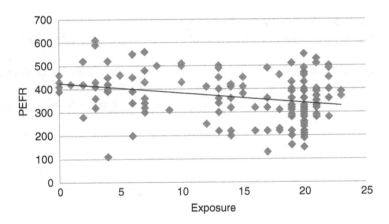

Figure 11.7   Trend line with negative slope.

The least squares regression line is:

$$y = -4.18458x + 424.5828$$

It appears that there may be a slight negative relationship between exposure to cotton dust and pulmonary effusion rates. This means that as the years of exposure increase, the pulmonary capacity decreases, albeit slowly. How much confidence can we have in this linear relationship?

Residual Plots

One method for measuring the predictive ability of a trend line is to look at a plot of residual values.

Figure 11.8 illustrates the residual plot for the baseball payroll data. At first glance, it appears from the plot that there is a large amount of error. To interpret the amount of error, and indirectly the confidence we have in our model equation, it is useful to compare the scale of the residuals to the scale of the number of wins in the original data. The residual values range from approximately +30 to −30, a range which is to some extent smaller than the range of wins (202–283). The fact that, in general, the error is smaller than the variability in the data reflects the fact that the regression equation is useful.

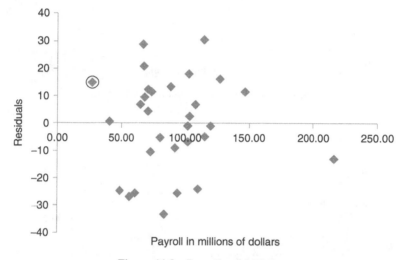

Figure 11.8   Payroll residual plot.

How to read the payroll residual plot:

The x-axis is the average annual payroll, that is, the independent or predictor variable. The y-axis is the residual between the predicted number of wins and the actual number of wins. Consider the highlighted point on the far left. The x-coordinate of this point is approximately \$27.07 million payroll for what was then called the *Florida Marlins* and is now called the *Miami Marlins*. Our regression equation predicted about 218 wins, and yet the Marlins won 233 games. Thus, the residual—actual wins–predicted wins—is 15, which is the y-coordinate of the point.

Figure 11.9, the Delta Wire training residual plot, has a residual error range of approximately −1600 to +1800. This seems high, but consider that the productivity range is from 70,000 to 90,000! The residual error is relatively small, and we can see this in the linear nature of the data shown in Figure 11.5.

At first glance, the residuals in the PEFR plot shown in Figure 11.9 appear to be similar to the previous plots. However, the residual error ranges from −300 to +210, and the actual data ranges from 110 to 610. The residual range is as large as the data range! This indicates that either the regression equation fit is not very good or there is not much of a relationship between the two variables. In this case, looking at the original data scatterplot shown in Figure 11.6, it does not appear that a strong relationship exists (Figure 11.10).

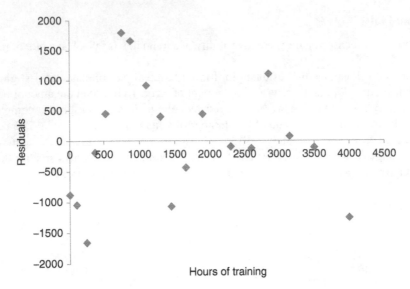

Figure 11.9   Delta Wire training residual plot.

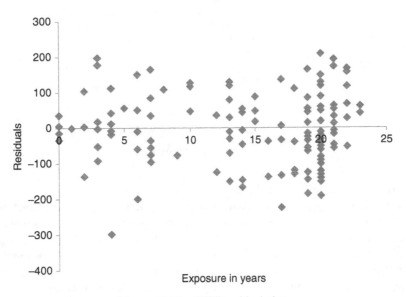

Figure 11.10   PEFR residual plot.

While the relationship is not strong, it still could be

1. statistically significant and
2. useful

The utility can be measured in terms of human health. Although respiratory function varies widely among individuals, an improvement in the overall average respiratory function is still very valuable. Most improvements in health conditions and treatments are not sweeping and revolutionary. If they help only 10% of the population, they are still very meaningful, but they will not necessarily show a dramatic statistical picture.

Note: There is probably some inherent difference among individuals with respect to PEFR, which adds noise to the picture. A plot of *change* in PEFR (which was not available), rather than PEFR itself, would control for the great differences among individuals and probably show a stronger relationship. See the paired comparisons section of Chapter 8 for a discussion of this issue.

11.4   INFERENCE FOR REGRESSION

Typically, slope coefficients in regressions are reported along with confidence intervals. This combination answers the question "how differently might this estimate of the relationship turn out if we selected additional samples from the same population?" We do not have lots of additional samples to examine in this case, and that will be true in most cases.

The next best thing is to take lots of bootstrap samples, that is, resamples with replacement from our original sample. The steps are as follows.

Resampling Procedure for a Confidence Interval (the Pulmonary Data)

Recall that the regression for the observed pulmonary suggested a negative relationship between pulmonary capacity and years of exposure data, with the regression equation:

$$y = -4.18458x + 424.5828$$

We want to establish a resampling confidence interval around both the constant and the slope coefficient. The procedure is as follows:

1. Place N slips of paper in a box, where $N =$ the original sample size. On each slip of paper, write down the variable values for a single case. In this case, a pair for one case comprises the values for exposure and PEFR. A slip of paper for one such case, for example, might read (0,390) where zero is the exposure and 390 is the PEFR.
2. Shuffle the papers in the box, draw a slip, and replace.
3. Repeat step 2 N times.
4. Perform a regression with PEFR as dependent variable and exposure as the independent variable and then record the coefficient and the constant.
5. Repeat steps 2–4, say, 1000 times.

You will end up with distributions of 1000 values for the coefficient and the constant, from which you can find appropriate percentiles to determine confidence intervals (e.g., the 5th and 95th percentiles for a 90% interval).

Using Resampling Stats with Excel (the Pulmonary Data, Continued)

Microsoft Excel has functions that you can enter in cells to calculate regression parameters (slope and intercept). It also has a regression routine in the data analysis set that reports traditional confidence interval values for the regression parameters, and more. These confidence interval calculations are formula-based on the assumption that the variables involved are Normally distributed.

A resampling approach, in contrast, makes the less restrictive assumption that the x–y pairs available for study were drawn from a much larger population of possible x–y pairs that is well represented by the sample at hand. To simulate this population and to use it to estimate confidence intervals, we draw randomly and with replacement from the set of x–y pairs—we bootstrap the cases as indicated earlier.

We saw earlier that the PEFR data set appears to be noisy in that it is difficult to estimate where the regression line should be. We are definitely interested in the reliability of the estimated slope and y-intercept. Figure 11.11 shows the first 17 rows of both the original data (columns A and B) and a bootstrap sample (columns D and E). The slope and the intercept for this single bootstrap sample can be seen in column G. The full procedure for 1000 resamples can be found in the textbook supplements.

	A	B	C	D	E	F	G	H
1	**Resampling and Regression**							
2								
3	Exposure:	PEFR:		**Resampled Rows as Units**				
4	0	390		14	320		-3.19578	<- Slope
5	0	410		20	260		405.0704	<- Intercept
6	0	430		19	310			
7	0	460		6	450			
8	1	420		22	450			
9	2	280		20	360			
10	2	420		21	510			
11	2	520		20	360			
12	3	610		19	320			
13	3	590		20	460			
14	3	430		19	160			
15	3	410		7	300			
16	3	360		20	360			
17	3	320		21	530			
18	4	110		15	450			
19	4	390		6	200			
20	4	400		15	380			

Formula bar: f_x =SLOPE(E4:E125,D4:D125)

Figure 11.11 Regression via resampling—revisiting the PEFR data with a single bootstrap sample (first 17 rows).

If you use the histogram function to analyze the Results sheet with sorted output values for the 1000 trials and check the cumulative output check box, you will see something like the output shown in Figure 11.12. It certainly indicates that the computed intercept in the Bin MidPt column has a wide range of values over the y-axis (PEFR) for the resampled data sets. Also notice the 90% confidence interval (the 5th and 95th percentile values).

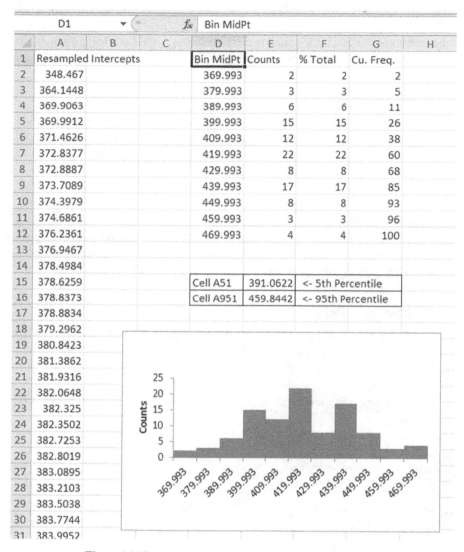

Figure 11.12   Analyzing PEFR regression intercept output.

Interpretation   The regression equation for the observed data from the Resampling Stats procedure was

$$y = -3.19578x + 405.0704$$

The 90% resampling confidence interval for the constant (intercept) goes from 391 to 460.

Try It Yourself 11.1

Try the simulation again using the slope value as a statistic of interest and obtain a resampling confidence interval for the slope.

Formula-Based Inference

Before resampling was available, confidence intervals were calculated using formulas based on the t-statistic that we saw earlier. Over-simplifying a bit, the interval is given by

$$b \pm t_{.05}SE$$

where b is the coefficient from the regression, $t_{.05}$ is the value of the t statistic at the 0.05 probability level, and SE is the standard error (SE) of b. The SE is calculated as

$$\frac{s}{\sqrt{\sum x^2}}$$

where s is the standard deviation of the residuals.

We present this for background only; analysts would nearly always perform regression using statistical software. Traditional simple output will show confidence intervals calculated automatically along the above lines.

Interpreting Software Output

All statistical software, and many other quantitative programs as well, perform regression. Let us look at the Excel output for an Analysis Toolpak regression and understand the various components (Figure 11.13):

	A	B	C	D	E	F	G	H	I
1	SUMMARY OUTPUT								
2									
3	*Regression Statistics*								
4	Multiple R	0.277021701							
5	R Square	0.076741023							
6	Adjusted R Square	0.069047198							
7	Standard Error	101.4381702							
8	Observations	122							
9									
10	ANOVA								
11		*df*	*SS*	*MS*	*F*	*Significance F*			
12	Regression	1	102633.2553	102633.3	9.974366	0.002008352			
13	Residual	120	1234764.286	10289.7					
14	Total	121	1337397.541						
15									
16		*Coefficients*	*Standard Error*	*t Stat*	*P-value*			*Lower 90.0%*	*Upper 90.0%*
17	Intercept	424.5828066	20.7960135	20.41655	7.51E-41			390.110276	459.0553371
18	X Variable 1	-4.18457649	1.324978603	-3.15822	0.002008			-6.3809285	-1.98822451

Figure 11.13   Excel's Analysis Toolpak regression output.

The output from the Analysis Toolpak regression macro is daunting at first glance. For our purposes, let us focus on the Coefficients section of the data described in the bottom section. Notice the "Intercept" and "X Variable 1" labels. Intercept refers to the y-intercept of the regression line (also called the constant) and X Variable 1 refers to the slope of the regression line. The Coefficients column supplies the values for the y-intercept and slope.

On the basis of the data shown in Figure 11.13, with coefficients rounded to two decimals, the resulting equation for the regression line (rounded to two decimal places) is

$$y = -4.18x + 424.58$$

The SE and t statistic are reported, as is a p-value and the lower and upper bounds of a 90% confidence interval. The SE and t statistic are both used in calculating the p-value and the confidence interval and so are redundant information. The p-value is used in determining whether the coefficient value is different from 0 to a statistically significant degree.

Note: In Excel you need to specify the degree of confidence for your interval—90%, 95%, and so on. The default confidence interval is 95%.

We ignored the confidence interval setting in the resampling procedure, because we were determining our own confidence interval via resampling, but if we use the Analysis Toolpak to determine the confidence intervals of the intercept and slope, we must set the appropriate confidence level in the dialog. The confidence level and resulting intervals are set at 90% as shown in Figure 11.13.

On the basis of the data from Figure 11.13, the 90% confidence interval for the intercept, rounded to two decimal places, is 390.11–459.06. The 90% confidence interval for the slope is −6.38 to −1.99.

At this point, the remainder of the regression output can be safely ignored.

11.5   EXERCISES

1. An online retailer wants to learn more about the success of its online advertising program. It reviews 3 weeks worth of sales of a variety of different products at different price ranges and analyzes the data in a regression. The variables involved are sales (Y) and ad costs (X) (each in $ 000). The following information is obtained:

```
intercept: coef. = 32 SE = 5.6
X1: coef. = 1.6 SE = 1.9
```

    (a) Write out the regression formula that the output shown earlier represents and interpret it fully in words.

    (b) What does "SE" stand for? What information does "SE = 1.9" give you, in a general sense?

    (c) For advertising expenditures of $10,000, what level of sales is predicted? How much confidence (a lot? a little?) do you have in the precision of this prediction?

2. A retail company wants to find out whether clickthroughs are a good substitute for sales in evaluating the effectiveness of an online ad. One clickthrough is one person clicking on an ad to learn more. Clickthroughs have the advantage of being much more plentiful than sales and accumulating much more quickly, allowing the firm to judge quickly whether an ad is effective. Here is some data on sales and clickthroughs for 13 ads (you can download an Excel workbook with the data clickthroughs.xls):

Clickthroughs	Sales
121	5
79	0
254	4
189	3
143	3
201	3
465	7
324	8
56	2
322	5
287	4
401	6
521	3

(a) Calculate (using a software program, if you wish) the correlation coefficient and explain how it can be used to assess whether clickthroughs are a good substitute for sales.

(b) Assume now that the company has determined that clickthroughs are, indeed, an adequate proxy (substitute) for sales. Determine (using linear regression and a software program) what the relationship is between clickthroughs and sales. What level of sales would you predict for 350 clickthroughs?

3. The following questions use this table, which can be downloaded here in Excel format in girls.xls. These data are the heights (in cm) of girls of different ages (in years). For the following questions, you may use software of your choice. (Our source is Siegel and Morgan, *Statistics and Data Analysis: An Introduction*, 2nd. ed., John Wiley and Sons, 1996. Their source is the 1980 World Almanac).

Height	Age
86.5	2
95.5	3
103.0	4
109.8	5
116.4	6
122.4	7
128.2	8
133.8	9
139.6	10
145.0	11

(a) Find the least squares regression equation for predicting height from age.

(b) Interpret the regression coefficients in terms of the growth of girls.

(c) Use the regression equation to predict the "height" of a 100-year-old "girl." Comment on the result.

(d) Plot the residuals from the regression against age. (You may plot standardized residuals if your software prefers those, and you may plot versus predicted height if your software likes to do that.) Do you see anything in the residual plot that was not obvious in the original scatterplot?

(e) Produce boxplots (same as box and whiskers plots) of x and y separately; you can use Box Sampler or StatCrunch. Comment on whether the distributions are mostly symmetric or asymmetric.

Answers to Questions

11.1 Extrapolation is always risky. In Figure 6.2, one might extrapolate that a $0.00 payroll would still result in 208 wins over three seasons. Similarly, a payroll of $800 million would result in 528 wins. Both cases are nonsensical. Major league ballplayers expect to be paid a salary, and there are only 486 regular season games in 3 years!

12

ANALYSIS OF VARIANCE—ANOVA

So far in hypothesis testing, we have been concerned mostly with single tests:

- Is humidifier A better than humidifier B?
- Which hospital error-reporting regime is better—no-fault or standard?
- Does providing additional explanation on the web reduce product returns?

We also looked at comparisons among multiple groups with categorical data. Now, we move on to comparisons among multiple groups for continuous numeric data. When we want to compare more than two groups at a time, we use a technique known as *analysis of variance*, or *ANOVA*. The technical details of ANOVA, and especially the hypothesis testing, will be of interest mainly to the *research* community. However, *data scientists* who analyze multigroup experiments will find the graphical exposition and the discussion of variance decomposition of benefit.

After completing this chapter, you should be able to

- compare multiple groups using boxplots,
- explain the problems involved with multiple testing,
- explain how the observations in a multigroup experiment can be decomposed into an overall average component, a treatment component, and a residual component,
- explain interaction, and include it in an ANOVA analysis,
- explain what factorial design is, what its advantages are, and what types of studies it is useful for,
- explain the role of blocking in experiments.

Introductory Statistics and Analytics: A Resampling Perspective, First Edition. Peter C. Bruce.
© 2015 John Wiley & Sons, Inc. Published 2015 by John Wiley & Sons, Inc.

12.1 COMPARING MORE THAN TWO GROUPS: ANOVA

Although the following data dates back to 1935, the purpose of the study—the amount of fat in our diet—is still a very current issue. In this study, investigators wanted to know how much fat doughnuts absorb while they are being fried. In particular, they wished to compare the absorption levels of four types of fat.

The investigators whipped up a batch of doughnut dough, split it into four doughnuts, and fried one doughnut in each type of oil (or fat). The results are given below.

```
Fat 1 164 grams
Fat 2 178 grams
Fat 3 175 grams
Fat 4 155 grams
```

If each fat is always absorbed to exactly the same degree, and there are no measurement errors, then the data mentioned above answer our question. Unfortunately, it is very likely that the quantity of fat absorbed varies from one batch to the next, even if we use just one fat. To assess this variability, we must repeat the experiment to see if we get exactly the same results. Our investigators did this. The second time around, they obtained the following results:

```
Fat 1 172 grams
Fat 2 191 grams
Fat 3 193 grams
Fat 4 166 grams
```

The results shown earlier are clearly not the same numbers when compared to what the investigators observed the first time around. For example, during the second experiment, Fat 3 was absorbed the most, whereas in the first experiment, Fat 2 was absorbed the most. A repetition of an experiment that allows us to assess variability is called a "*replication.*"

Definition: Replication

A replication is a single repetition of an experiment or a procedure with all elements remaining unchanged.

Actually, our investigators made six replications, using just one large batch of dough.

Question 12.1

Discussion Question

What are the advantages and disadvantages of using one big batch of dough, rather than a separate batch for each replication?

In at least one replication, Fats 1, 2, and 3 each had, at one point, the highest levels of absorption. There were also replications in which Fats 1 and 4 had the lowest levels of absorption. Now, our results are not so clear-cut. Let us look at the complete data set in Table 12.1.

TABLE 12.1   Fat absorption data (grams)

	Fat 1	Fat 2	Fat 3	Fat 4
Replication 1	164	178	175	155
Replication 2	172	191	193	166
Replication 3	168	197	178	149
Replication 4	177	182	171	164
Replication 5	156	185	163	170
Replication 6	195	177	176	168

Of course, it is always important to graph our data in addition to examining summary statistics. Figure 12.1 shows dotplots of the data, and Figure 12.2 shows boxplots of the data.

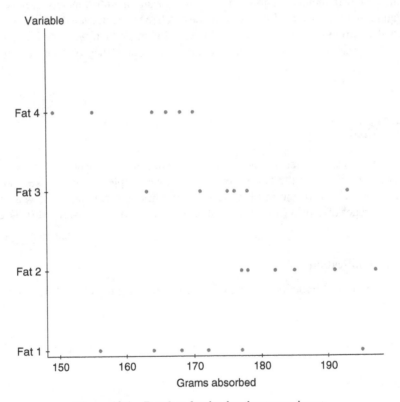

Figure 12.1   Dotplots for the doughnut experiment.

It looks like the groups differ, but it also looks like there is a lot of overlap among the groups. We need some criterion for deciding just how different the samples must be before we will say there is a real difference in the population. In other words, we need some sort of hypothesis test.

What statistic will we use in our hypothesis test? There are several choices, but the average amount of fat absorbed is probably of primary interest, so we will compare means, as shown in Table 12.2.

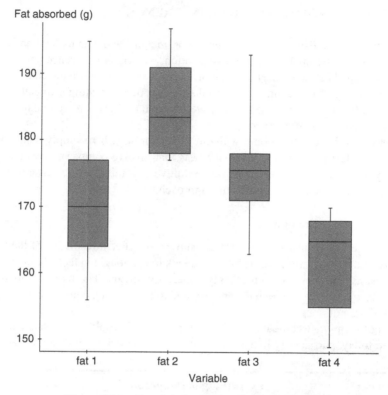

Figure 12.2   Boxplots for the doughnut experiment.

TABLE 12.2   Comparing means

	Fat 1	Fat 2	Fat 3	Fat 4
Replication 1	164	178	175	155
Replication 2	172	191	193	166
Replication 3	168	197	178	149
Replication 4	177	182	171	164
Replication 5	156	185	163	170
Replication 6	195	177	176	168
Average	172	185	176	162

Now, we have a conundrum. When we were comparing just two groups, it was a simple matter; we merely looked at the difference between the means of each group. With four means, there are six possible comparisons between fat types.

Fat 1 compared to Fat 2

Fat 1 compared to Fat 3

Fat 1 compared to Fat 4

Fat 2 compared to Fat 3

Fat 2 compared to Fat 4

Fat 3 compared to Fat 4

12.2   THE PROBLEM OF MULTIPLE INFERENCE

One common abuse of data analysis is to gather or find data and then look up and down and left and right in an attempt to find some association, pattern, or difference between groups that is interesting. The more questions you investigate, the greater are the chances that you will find something "significant." The related sin in hypothesis testing is to gather data first and then test all kinds of hypotheses until we find one that gives a significant result. We often refer to this sin as *data snooping*.

One option we have at this point, with our doughnut data, is to simply conduct six different two-sample *t*-tests (or six different *t*-tests that involve comparing two independent means, as you learned about in Chapter 8). Intuitively, it might make sense to you to do this, but there are some problems with this approach.

The False Discovery Problem

Because each test or comparison will falsely show a significant result 5% of the time, you can usually find such an erroneous result if you look long enough; in fact, about one in every 20 comparisons you make will mistakenly appear significant. The more comparisons we make, the more prone we are to make an error and to claim that a comparison is significant when it really is not.

This is not simply a technical issue. It is a fundamental problem with data analysis as it is practiced today, particularly in the health arena, and there are significant costs.

"Everything is Dangerous"

"Everything is dangerous" was the title of a lecture that statistician Stanley Young gave at a number of US government agencies in the first decade of the twenty-first century. He was referring to published reports about associations between certain things and human health, such as coffee and pancreatic cancer, type A personality and heart attacks, and reserpine and breast cancer.

Young and a colleague, Alan Karr, expanded on this theme in a September 2011 article in *Significance* magazine. They examined 52 claims of health benefits reported in 12 peer-reviewed papers. All these papers reported observational studies rather than experiments (i.e., they were reviews of existing data). One of the observational studies claimed protection from heart attack afforded by vitamin E, which we mentioned at the beginning of this book.

Of course, there are hundreds—thousands—of claims every year about substances or treatments represented as helpful or harmful to health. What was special about these 52 claims was that they were later tested in rigorously controlled randomized trials. Amazingly, *not one* of the 52 claims was verified, and the trials produced opposite results in five of the cases.

The most likely reason for the failure to validate these claims in experiments is that they were the result of data snooping in the first place. Data snooping is an extensive search through data, testing a variety of possibilities until something interesting is found.

A conservative approach that avoids such problems is to state any hypotheses to be tested in advance. If a researcher cannot state a definite hypothesis, then it is probably too early in the study to test anything, and more data is needed to develop at least one hypothesis.

If you have more than one hypothesis to test, then you must adopt a more stringent standard for any one of them to be declared "significant." A simple, though conservative, approach is to divide the desired alpha by the number of tests. For example, if you have five hypotheses to test, and the traditional alpha (i.e., significance level) you require for a single hypothesis test is 0.05, then the equivalent alpha for each of five hypothesis tests is 0.01 (or 0.05/5). There are more complex schemes for alpha adjustment, but they are beyond the scope of this course.

For now, we will not worry so much about conducting multiple hypothesis tests but will instead focus on a procedure that allows us to examine multiple groups without having to conduct multiple tests. We conduct what is called an *analysis of variance* (ANOVA) in order to avoid having to conduct many individual t-tests. If we use ANOVA, we can examine all groups within a single test. If the results of this test are significant, this means there is some significant difference among the group means. The distinction between the approaches can best be seen in the null models that are being tested:

ANOVA

Tests the null model that all groups could be drawn from the same box.

Multiple *t*-Tests

Each one tests the null model that the two groups in the test could be drawn from the same box.

12.3   A SINGLE TEST

Our point in discussing the problem of multiple inference is simple. If we return to the doughnut problem, instead of worrying about all the different comparisons between individual fats we could make, in addition to our increased chances of making errors were we to make so many comparisons, we can do a single overall test that addresses the question, "Could all the fats have the same absorption property?"

If all the fats have the same absorption property, this means they may come from the same population. Our statistical hypothesis to be tested then becomes, "Could all the fat absorption values come from the same population?" The null hypothesis is that they do, and that any differences between fats are just due to chance. We can test this hypothesis by placing all the values in a box and then drawing out groups of six. Before we do that, we need a way to measure the extent to which each mean varies from each of the other means. As you have seen already, there are different ways to measure variation, such as the range, sum of absolute deviations, variance, standard deviation, and interquartile range.

We could use any of the above measures, but we will choose the variance. A simple way of thinking about ANOVA is that we are comparing two measures of variance. We are examining the variability in our response, or our dependent measure, that is explained by the independent variable, or by the treatment (in this example, by the type of fat), and we are comparing this to the variability in the response that is explained by "residual error," or by other factors that are beyond the control of the investigator. We call this "residual error" because it is basically the variability that is left over once we account for the effect of our independent variable. Ideally, we want more of the variability in the response to be explained by treatment, or by the independent variable, and less to be explained by "residual error."

12.4   COMPONENTS OF VARIANCE

George Cobb, in his text *Introduction to Design and Analysis of Experiments* (2008), presents a useful approach to ANOVA that "decomposes" individual observations into components. See that text for a more complete discussion of the following material.

In an experimental design such as the one we are focusing on, each observed data value in the data set can be thought of as a combination of three components:

Observed value = Grand average + Treatment effect + Residual error

Sometimes, this ANOVA model is written in symbols rather than in words:

$$Y_{ij} = \mu + \alpha_i + e_{ij}$$

In the model described earlier, Y_{ij} is used to denote the individual observation. The grand average is denoted as μ, the treatment effect as α_i, and residual error as e_{ij}.

As we talk about this model and what each part represents, we will refer to the doughnut data as a concrete example. You can see this data again in Table 12.3, along with the means for each fat and the grand average.

TABLE 12.3   Doughnut data with group means and grand average

Fat 1	Fat 2	Fat 3	Fat 4
164	178	175	155
172	191	193	166
168	197	178	149
177	182	171	164
156	185	163	170
195	177	176	168
Average $(\bar{x}_1) = 172$	Average $(\bar{x}_2) = 185$	Average $(\bar{x}_3) = 176$	Average $(\bar{x}_3) = 162$
			Grand average = 173.75

The goal of ANOVA is to estimate the amount of variability in our dependent (response) variable that is due to the independent variable, as opposed to variability due to other factors, such as individual differences among cases or participants. When you use ANOVA, you hope to find that most of the variability in your response is due to the independent variable because you want to be able to demonstrate that the independent variable has a particular effect on the dependent variable. To perform an ANOVA, we need a way of estimating just how much of the variability in our dependent variable can be attributed to the independent variable and just how much can be attributed to other factors. As noted earlier, the variability not attributed to the independent variable is referred to as *residual error*, or sometimes just error. It is the variability that is left over that cannot be explained by the independent variable.

Remember, again, that each observed value in our data set can be considered a sum of three different parts:

Observed value = Grand average + Treatment effect + Residual error

Consider an assembly line metaphor in order to help explain the model described earlier. Think of something like a car and how the car is built. The car might be built using an assembly line. All cars start out the same in that each car has a basic frame, or foundation, upon which we can build. The car slowly moves down the assembly line, and different pieces are added to the foundation along the way. In the end, the sum of all pieces forms the car itself.

Observed values in a data set can be considered sums of parts just as well.

1. Start with grand average (173.75 for doughnut data).
2. Add treatment effect, which might be positive or negative (independent variable = fat type).
3. Add residual error, which might be positive or negative.

For any observed data value within a data set, we can break it down into the grand average, the treatment effect, and the residual error. We call this a "decomposition" with our doughnut data. If you understand what we are doing here, it will hopefully make it easier to see what the numbers we get in our statistical analysis actually tell us.

Decomposition: The Factor Diagram

When we perform a decomposition, we find the means for each of our treatment groups and the grand average (or grand mean). As mentioned earlier, this information is included in Table 12.3. We can take this information and put together a factor diagram. This will allow us to better visualize how each observation in our data set is a sum of different parts. It will also give us the information we need to construct an ANOVA table. The factor diagram is below (in Figure 12.3), and following the diagram is a discussion of how the different components in the diagram were calculated.

The Observation: In this part of our diagram, we are simply presenting the individual observations in our data set. Note that each column in the observation portion of the diagram refers to a different group in our study (column 1 is Fat 1, column 2 is Fat 2, column 3 is Fat 3, and column 4 is Fat 4).

The grand average: We know that in our example, the average of all the observations is 173.75. This is our grand average. Remember that in our ANOVA model, the grand average is the one thing that all observations have in common as all observations are involved in computing the grand average. This is our baseline or where the observation begins. As the grand average is the same for everyone, we have just repeated the same value 24 times in the grand average portion of the diagram.

The treatment effect: For each group, we can find a group (treatment) mean, and learn how much each group diverges from the grand average. This divergence is shown in the treatment effect portion of the diagram. In some groups, we see the treatment results in increased values of the dependent variable, on average. In other groups, the treatment results in decreased values of the dependent variable. Column 1 of the treatment effect portion of the factor diagram shows us how the mean for Fat 1 deviates from the grand average. We get the value of −1.75 by taking the mean from Fat 1 (or 172) and subtracting

Observation

164	178	175	155
172	191	193	166
168	197	178	149
177	182	171	164
156	185	163	170
195	177	176	168

=

Grand average

173.75	173.75	173.75	173.75
173.75	173.75	173.75	173.75
173.75	173.75	173.75	173.75
173.75	173.75	173.75	173.75
173.75	173.75	173.75	173.75
173.75	173.75	173.75	173.75

+

Treatement effect

-1.75	11.25	2.25	-11.75
-1.75	11.25	2.25	-11.75
-1.75	11.25	2.25	-11.75
-1.75	11.25	2.25	-11.75
-1.75	11.25	2.25	-11.75
-1.75	11.25	2.25	-11.75

+

Residual error

-8	-7	-1	-7
0	6	17	4
-4	12	2	-13
5	-3	-5	2
-16	0	-13	8
23	-8	0	6

Figure 12.3 Factor diagram of doughnut data example.

the grand average (or 173.75) from it, to give us $172 - 173.75 = -1.75$. The second column shows how the mean from Fat 2 (or 185) deviates from the grand average, and this gives us $185 - 173.75 = 11.25$. Hopefully, you can see how you would apply this same logic to determine the treatment effects for Fats 3 and 4. Note that if you now look across each row in the treatment effect portion of the diagram and add all the deviations (or treatment effects) together, they should sum or add to 0. In row 1, for example, we have $-1.75 + 11.25 + 2.25 + (-11.75) = 0$. The treatment effects sum to 0 because some group means are above the grand average (or sometimes even equal to the grand average) and some are below the grand average. If the estimated treatment effects do not sum to 0, you know you have made a calculation error when constructing the factor diagram.

The residual error: Once we know the grand average and the treatment effect, what is left over is the residual error. Ideally, we want our residual errors to be small. We want to control all we possibly can so we can conclude that it is the treatment, or the independent variable, that led to changes in the dependent measure, not other variables we did not control. To estimate the residual error for each observation, we simply take the observation and subtract the group mean from it. Our residual errors tell us how individuals or cases within each group cluster around their group mean. So, for example, remember that column 1 contains all doughnuts fried in Fat 1. The first case in that group had had a value of 164. If we take 164 and subtract 172 from it (as 172 is the mean for Fat 1), we get -8. We must do this now for each observation in our data set (i.e., take the observation and subtract from it the mean from the group it is in). If we add all the deviations in a single column together, they should sum to 0. In column 1, for example, we would have $-8 + 0 + (-4) + 5 + (-16) + 23 = 0$.

Constructing the ANOVA Table

Once we have decomposed the data, we can use the factor diagram to help us construct an ANOVA table and conduct the ANOVA. To construct our table, we need to determine degrees of freedom (df), sums of squares, mean squares, and an F-statistic. Let us first examine how we would determine the appropriate degrees of freedom for each source of variability in our analysis.

The df is the number of observations that are theoretically free to vary once some parameter(s) describing the data is set. For example, suppose you have two observations and, the variance is fixed at 0. Once you know that one of the observations is equal to 2, the other observation is not free to vary, so it must also be equal to 2 in order to produce variance $= 0$.

For each part, or section, of our factor diagram, we can determine a corresponding value for the degrees of freedom.

- For the grand average section, all values are, by definition, the same (as the grand average is the same for everyone). Once we know the grand average and place it in one cell of the grand average portion of the factor diagram, we know all other cells will have the same value.
- So, we only have one degree of freedom associated with the grand average. For the treatment effect section of the factor diagram, recall that the treatment effects must sum to 0 within each row. In this example, we have four treatment groups. Once we know three of our estimated treatment effects, the fourth must be that value that causes all treatment effects to sum to zero. So, in this example, within the treatment effect section, only three cells are theoretically free to be anything. Once we *know the first* value in the first three columns, we know what the first value in the fourth column must be (as the values across rows must sum to 0). We also know that all values in a

single column must be the same as the first value in that column (as all members of the same group get the same treatment effect), so once we know what the first value in the column is, all other values in that column are fixed. There is also a formula that we can use to determine the degrees of freedom due to treatment. It is the number of groups (denoted sometimes as "k") minus 1. In our example, we have $k - 1 = 4 - 1 = 3$.

- For the *Residual Error* section, we know that each column gives the deviations around the group mean. For each column, once we know all but one of the deviations, the last one is fixed (as the sum of the deviations must be 0). In this particular example, within each column, we have $6 - 1$ degrees of freedom, or 5. Thus, the degrees of freedom associated with residual error would be 20 (or 5×4, as we have four groups). We can also use the formula $N - k$ to determine the degrees of freedom for residual error, where "N" is the total sample size (in this case 24) and "k" is the number of groups (in this case 4). We can see that $24 - 4 = 20$.

Before we extend this discussion and describe the other components of the ANOVA table, it might be helpful to see the analysis from the resampling perspective.

Resampling Procedure

Keep in mind our original question: Is the observed variance among group means greater than what might occur by chance alone?

Before we go into details about the resampling procedure, let us look at our data in a different way.

Recall from your previous textbook readings that there are two possible variance measures—one used to describe a population and another used when working with samples. In particular, Section 1.6 in the textbook includes a discussion of variance, and it would be good here to review that section if you need to.

We will use the sample variance for the doughnut experiment. The formula for the sample variance is shown below:

$$\text{Sample variance} = s^2 = \frac{\sum (x - \bar{x})^2}{n - 1}$$

This formula, however, is for the variation between *individual observations* and the sample mean. For our doughnut experiment, we need to adapt it to measure the variation between *group means* and the overall mean or grand average. This measure is shown below, where x-bar is the average for each of the k groups (and keep in mind that k refers to the total number of groups).

$$\text{Variance among group means} = \frac{\sum (\text{grand average} - \bar{x})^2}{k - 1}$$

Table 12.4 shows the variances among group means for the experiment. Hopefully, you can see how some of these numbers relate back to the decomposition we just did!

"Residual" in the table shown earlier takes on a slightly different meaning than the way we were using "residual" when talking about decomposition; here, "residual" refers to the difference between the group mean and the grand average.

If we wanted to use a resampling procedure, we could conduct our analysis as follows:

TABLE 12.4   Variance of group means

Fat Type	Group Mean	Residual	Res. sq.
1	172	−1.75	3.06
2	185	11.25	126.56
3	176	2.25	5.06
4	162	−11.75	138.06
	Grand mean 173.75		Total 272.75

Variance $= 272.75/(k-1) = 272.75/3 = 90.92$.

1. Place all 24 fat absorption values in a box.
2. Shuffle the box, draw* four groups of six values each (i.e., four groups, each with six values).
3. Find the mean of each of the four groups.
4. Calculate the variance among the means and record it.
5. Repeat steps 2 through 4 many more times (say up to 1000 times).
6. Determine what proportion of the 1000 trials produces a variance ≥ 90.92. This is the *p*-value.

*When you combine two or more groups in a single box, drawing from that box for a hypothesis test can be done with or without replacement. The conceptual frameworks and statistical properties differ a bit in ways that are beyond the scope of this text, but both are valid procedures.

A specific software procedure for this example can be found in the textbook supplements.

If the trials regularly produce a variance in excess of the observed value of 90.92, this indicates that chance is a reasonable explanation for the differences among the groups.

The largest 10 results out of 1000 trials produced by a Resampling Stats procedure, and ordered from largest to smallest, are given below

103.3796
101.8611
98.15741
95.82407
94.63889
94.08333
90.15741
89.41667
87.80556
87.65741

You can see that only six of the 1000 trials produced a variance as great as 90.92, for an estimated *p*-value of 0.006. We therefore conclude that chance is probably *not* responsible for the differences among the fats with respect to the absorption in the doughnuts, and those differences are statistically significant.

As shown in Figure 12.4, the distribution of the variances of the 1000 trials is skewed to the right and truncated at zero. Variances less than zero are not arithmetically possible.

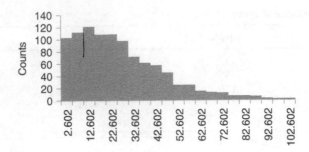

Figure 12.4   Frequency histogram of resampled variances from doughnut data problem.

Inference Using the ANOVA Table

We can also determine whether the differences among the fats are significantly different by formula, using the numbers in the ANOVA table. Instead of looking simply at the variance among the group means, we will look at the *ratio* of two measures of variation, one for the treatment and the other for the residual. This allows us to use a single comparison table for all ANOVA problems.

Let us return now to our decomposition example. If you recall, we began our discussion about constructing the ANOVA table by illustrating ways in which we could determine degrees of freedom (df). To complete our ANOVA table, we also need to obtain the sum of squares, the mean squares, the *F*-statistic, and a *p*-value.

Why Squared Deviations?

We square deviations before summing them because, in the case of the treatment effect and the residual error, if we simply added values together in the factor diagram without squaring them, the sum would always end up being zero and all sense of variation would be lost. This again is because the treatment effects represent how each group mean deviates from the grand mean (or grand average), and the residual errors represent how each data value deviates from the mean of the group it belongs to. (We could take absolute values of the deviations, but squared values are easier to deal with mathematically in calculations.)

We can figure out what is called the *sum of squared deviations* (or sum of squares, or SS) associated with each part of our model (grand average, treatment effect, and residual error) by squaring all values in the corresponding section of our factor diagram. The sum of squares is basically a measure of overall variability. For example, to find the sum of squares for the grand average, we take 173.75 and square it (to get 30,189.0625). Then, we would multiply this by 24 (or square the value of 173.75 a total of 24 times and add all the squared values together). This would give us the sum of squared deviations for the grand average. We would do the same thing for our treatment effects section (e.g., we would square each cell in the box and add all the squared values together) and for our residual error section. If you do this, you should get:

- Sum of squares (grand average) = 72,4537.5.
- Sum of squares (treatment effect) = 1636.5.
- Sum of squares (residual error) = 2018.

We next need to find averages of the sum of squares due to treatment and the sum of squares due to residual error. This is because the sum of squares gives us a measure of overall variability, but we need a way of taking into account sample size and total number of treatment groups. These averages are commonly called *mean squares* (or MS). We have mean squares for treatment and mean squares for residual error. We obtain the mean squares for treatment by taking the sum of squares for the treatment effect (or 1636.5) and dividing it by the degrees of freedom for the treatment effect (or 3). This gives us $1636.5/3 = 545.5$. We can also find the mean square for our residual error by taking the sum of squares for the residual error (or 2018) and then dividing it by the degrees of freedom for the residual error (or 20). This gives us $2018/20 = 100.9$. We do not worry at this point about finding a mean square for the grand average because we do not need that to calculate our F-statistic, and the F-statistic is what we use to determine if there is a significant difference among our treatment groups.

We get our *F-statistic* by dividing the mean square for our treatment effect by the mean square for our residual error. In this example, we get $F = 545.5/100.9 = 5.41$.

The formula for the F-statistic is:

$$F = \frac{MS_{\text{Treatment}}}{MS_{\text{Error}}}$$

We want our F-statistic to be significantly greater than 1 because we want more of the variability in the response (the dependent variable) to be explained by treatment (the independent variable) and not by residual error. If more variability is explained by treatment, this means there are differences among group (or treatment) means, and it is more support AGAINST the null hypothesis. When we conduct an ANOVA by means of resampling, or by using statistical software, we will obtain a p-value, and the p-value will help us determine if our F-statistic is likely or unlikely to have occurred simply by chance alone.

Now that you have a basic understanding of the ANOVA model, let us see how we can put the earlier mentioned pieces together into an ANOVA table.

In Figure 12.5, you will see the ANOVA table that was created based on our decomposition of the data. Underneath this table, you will see the ANOVA table that came from analyzing the data using StatCrunch. Note that when we decompose data by hand, we had to look up the F-statistic in a table to find an approximate p-value; the p-value is computed for us automatically when we use software. Also, depending on which software package you use, you may not see any information related to the grand average in the ANOVA table. This is okay as we do not need this information to help us calculate our F-statistic.

Source of Variability	SS	df	MS	F
Grand average	72, 4537.5	1		
Treatment	1636.5	3	545.5	5,4117.13
Residual error	2018	20	100.9	
Total	728, 192	24		

This example of decomposition was presented to help you get a conceptual understanding of what is happening within an ANOVA. We want you to feel comfortable using software to do this analysis as things get more complicated if you have larger samples (or if you have unequal numbers of cases per group). You can use resampling procedures

Analysis of variance results:
Data stored in separate columns.

Column means

Column	n	Mean	Std. error
Fat 1	6	172	5.4467115
Fat 2	6	185	3.172801
Fat 3	6	176	4.0331955
Fat 4	6	162	3.3565855

ANOVA table

Source	df	SS	MS	F-Stat	P-Value
Treatments	3	1636.5	545.5	5.406343	0.0069
Error	20	2018	100.9		
Total	23	3654.5			

Figure 12.5   ANOVA table from decomposition contrasted with ANOVA table from StatCrunch.

(which you will need to construct using a resampling program or a scripting language), or you can use the formula approach (which is built in as a standard procedure in *standard statistical software*). More details about using software can be found in the textbook supplements.

The *F*-Distribution

We have seen earlier that W. S. Gossett devised a formula for the *t-distribution* that approximates the distribution of a sample mean and the difference between two sample means. The *F* statistic distribution is an analogous distribution for the ratio of treatment and error variances.

To determine if we have a significant *F*-statistic (and to estimate our *p*-value), we compare our calculated *F*-statistic to a standard distribution of variance ratios. To do this, we need a family of standard distributions of these ratios. This family is called the "*F-distribution*," after Sir Ronald Fisher, who originally proposed the formal analysis of variance. Each member of this family is characterized by two separate values for degrees of freedom.

As you learned from our discussion of decomposition, df for the numerator (or for our treatment effect) is given as $= k - 1$, where $k =$ the number of groups. Degrees of freedom for the denominator (or for residual error) is given as $= k(n - 1)$, where k is again the number of groups and $n =$ the number of values in each sample. Note that in cases where there are an unequal number of cases per sample, the more general $(N - k)$, where N is the total sample size across all groups, can be used to find the degrees of freedom from the denominator.

As noted earlier, based on our decomposition of the data, the degrees of freedom for the numerator is 3 and the degrees of freedom for the denominator is 20. We can thus write this as $F_{(3,20)}$, and this refers to the F-distribution for four groups of six values each.

You can compare the observed F-value (5.406) to a critical F-value by looking it up in a table in a textbook, or you can use a web calculator to determine the area to the right of the observed F-statistic. The two methods are equivalent. Figure 12.6 shows what this would look like using a web calculator.

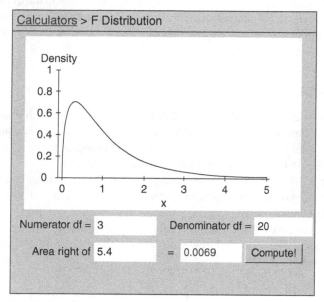

Figure 12.6 *F*-Statistic web calculator with graph (Source: http://www.stat.tamu.edu/~west/applets/fdemo.html).

From Figure 12.6, we can see a *p*-value of 0.0069. This matches closely with the *p*-value we saw earlier in our StatCrunch output and in the results from using a resampling procedure.

Note that the statistic that we used in our resampling procedure was the variance among group means. We could also have used the F-statistic itself as our resampling statistic, calculating it during each resample and comparing the resampled distribution to the observed value. Fisher, like Gossett, simply found the mathematical equivalent of the resampled distribution and tabulated it.

Different-Sized Groups

So far, we have only dealt with situations where the groups are all equally sized. What about situations where the group sizes are different? In practice, this is a common situation a researcher must understand how to deal with.

Resampling Method   When group sizes differ, only step 2 in the resampling method outlined following Table 12.4 changes. Instead of shuffling the box and drawing four groups of six values each, as before in step 2, we now shuffle the box and draw four groups with the same number of values as in each original group. So, as an example, suppose one group has three values, the second has four values, the third has six values, and a fourth group has seven values. For this case, we shuffle the box and draw three values for the first group, four for the second, and so forth.

Formula Method   In contrast to the situation with the resampling method, the calculation of the F-statistic described earlier works only when the groups are all of the same size. If the groups are not of the same size, the computations are messier. However, in all likelihood you will be doing your analysis with software and most statistical software can easily handle different group sizes, and if you use a software that can handle different-sized groups, the numbers you get in your printout (e.g., numbers for "SS," "MS," and "F") will have exactly the same meaning as described earlier.

Caveats and Assumptions

ANOVA may be used with multigroup controlled experiments or with multigroup observational studies where no treatment is involved. For experiments, treatments must be randomly assigned to subjects. For observational studies, sample groups must be independently and randomly selected from a much larger population of interest.

In addition, for the formula-based approach, which includes most software applications, we assume that our groups are similar in terms of variability (e.g., we assume the largest group standard deviation is no bigger than twice the smallest group standard deviation), and we assume that the outcome variable is Normally distributed. These requirements can be checked by comparing standard deviations among the groups by graphical analysis or by examining residuals.

12.5   TWO-WAY ANOVA

Running an experiment is expensive, and it is often cost-effective to study more than one thing at a time. Consider the following experiment to test how soldering is affected by the addition of antimony to the solder, which lowers costs, and by the joint cooling method, which cools the joint in a reasonable amount of time for manufacturing.

Solder is a metal alloy with a low melting point. It is used to bond metals carrying an electrical current. This is how it works: Suppose we wish to connect two wires. We can just twist them together, but then they can easily come loose or corrode. Instead, we add solder at the wire joint while applying heat, thereby melting the solder. When the soldered joint cools, the solder solidifies, holding the wires together and protecting them from dirt and corrosion. Being a metal, the solder conducts electricity. Therefore, it both bonds the two wires together and keeps the circuit intact.

The experiment that we will examine now was conducted and reported by Tomlinson and Cooper in "Fracture mechanism in brass/Sn–Pb–Sb solder joints and the effect of production variables on the joint strength," (measured as MN/mm^2 where N = Newtons, a measure of pressure) *Journal of Materials Science*, Vol. 21, No.5, May 1986. Our source for the data is Mendenhall and Sincich, *A Second Course in Statistics: Regression Analysis* (5th ed.), Prentice Hall, Upper Saddle River, NJ, 1996.

The amount of antimony used was one variable, and it was studied at four levels—0%, 3%, 5%, and 10%. The other variable was the cooling method for the solder joints. Large-scale soldering operations generate large amounts of heat. Some means of forcibly cooling the joints is needed to achieve sufficient cooling in a reasonable amount of time. Four cooling methods were used. Forty-eight wire joints were prepared, and three were randomly assigned to each of the $4 \times 4 = 16$ treatment combinations.

The first few rows of the data are shown in Table 12.5 and the entire data set can be downloaded at the book website.

Cooling Method	Joint Strength
1	17.6
1	19.5
1	18.3
2	20
2	24.3
2	21.9
3	18.3
3	19.8
3	22.9
4	19.4
4	19.8
4	20.3
1	18.6
1	19.5
1	19

TABLE 12.5   Initial rows of antimony data

Antimony	Cooling Method	Joint Strength
0	1	17.6
0	1	19.5
0	1	18.3
0	2	20
0	2	24.3
0	2	21.9
0	3	18.3
0	3	19.8
0	3	22.9
0	4	19.4
0	4	19.8
0	4	20.3
3	1	18.6
3	1	19.5
3	1	19

Interaction

So far, we have considered variables under study that appear to be independent of one another. For instance, consider the antimony data that we have examined earlier. It was observed that antimony had an effect on solder strength, and the cooling method had an effect on solder strength, but the interaction between the two variables was not significant. This suggests that the antimony effect does not depend on the cooling method chosen.

This is not always the case. Sometimes, there is an interaction between variables, meaning that the effect of one variable differs, depending on the presence or the level of another variable.

Consider grapefruit and quinidine. Alternative medicine practitioners advocate consuming grapefruit to minimize the symptoms of malaria. Quinidine is also prescribed by traditional doctors for malaria. So a study of the two variables might show that each, by itself, is associated with diminished malarial symptoms. However, there is an interaction between grapefruit and quinidine. Grapefruit affects certain enzymes of the stomach in ways that inhibit the absorption of quinidine. Therefore, if you take quinidine, adding grapefruit will probably exacerbate malarial symptoms, rather than adding further relief.

As we go from one-way ANOVA to two-way ANOVA, we will explicitly account for possible interaction.

Components of Variance

Just like with the one-way ANOVA example, observed values in an experimental design that involve two factors, or independent variables, can be thought of as a combination of several components:

Observed value = Grand average + Treatment 1 effect + Treatment 2 effect

+ Interaction effect + Residual Error

Sometimes, this ANOVA model is written in symbols rather than in words:

$$Y_{ij} = \mu + \alpha_i + \beta_i + (\alpha_i \beta_i) + e_{ij}$$

In the above model, Y_{ij} is used to denote the individual observation. The grand average is denoted as μ, the effect of the first independent variable or treatment as α_i, the effect of the second independent or treatment as β_i, the interaction between the independent variables or treatments as $\alpha_i \beta_i$, and the residual error as e_{ij}.

The term $\alpha_i \beta_i$ looks like the product of two variables but it is not—it means the effect due to the two variables together, above (or below) the sum of the effects of two variables taken individually.

Just as we did within the context of a one-way ANOVA, we can decompose the antimony data from Table 12.5 to better appreciate what the model described earlier represents, in addition to better understanding where the numbers in the ANOVA table come from.

Decomposition

The complete factor diagram for the antimony data set is shown in Figure 12.7.

Just as we did within the context of a one-way ANOVA, we begin by finding the *grand average*. Each "slot" in the grand average portion of our factor diagram will contain the value of the grand average. In the antimony data set, the grand average is approximately 19.55.

Observation

17.6	18.6	22.3	15.2
19.5	19.5	19.5	17.1
18.3	19	20.5	16.6
20	20	20.9	16.4
24.3	20.9	22.9	19
21.9	20.4	20.6	18.1
18.3	21.7	22.9	15.8
19.8	22.9	19.7	17.3
22.9	22.1	21.6	17.1
19.4	19	19.6	16.4
19.8	20.9	16.4	17.6
20.3	19.9	20.5	17.6

+

Grand average

19.55	19.55	19.55	19.55
19.55	19.55	19.55	19.55
19.55	19.55	19.55	19.55
19.55	19.55	19.55	19.55
19.55	19.55	19.55	19.55
19.55	19.55	19.55	19.55
19.55	19.55	19.55	19.55
19.55	19.55	19.55	19.55
19.55	19.55	19.55	19.55
19.55	19.55	19.55	19.55
19.55	19.55	19.55	19.55
19.55	19.55	19.55	19.55

+

Antimony

0.62	0.86	1.07	-2.53
0.62	0.86	1.07	-2.53
0.62	0.86	1.07	-2.53
0.62	0.86	1.07	-2.53
0.62	0.86	1.07	-2.53
0.62	0.86	1.07	-2.53
0.62	0.86	1.07	-2.53
0.62	0.86	1.07	-2.53
0.62	0.86	1.07	-2.53
0.62	0.86	1.07	-2.53
0.62	0.86	1.07	-2.53
0.62	0.86	1.07	-2.53

+

Method

-0.91	-0.91	-0.91	-0.91
-0.91	-0.91	-0.91	-0.91
-0.91	-0.91	-0.91	-0.91
0.9	0.9	0.9	0.9
0.9	0.9	0.9	0.9
0.9	0.9	0.9	0.9
0.63	0.63	0.63	0.63
0.63	0.63	0.63	0.63
0.63	0.63	0.63	0.63
-0.6	-0.6	-0.6	-0.6
-0.6	-0.6	-0.6	-0.6
-0.6	-0.6	-0.6	-0.6

+

Interaction

-0.79	-0.47	1.06	0.19
-0.79	-0.47	1.06	0.19
-0.79	-0.47	1.06	0.19
1	-0.88	-0.05	-0.09
1	-0.88	-0.05	-0.09
1	-0.88	-0.05	-0.09
-0.47	1.19	0.15	-0.92
-0.47	1.19	0.15	-0.92
-0.47	1.19	0.15	-0.92
0.26	-0.88	-1.19	0.78
0.26	-0.88	-1.19	0.78
0.26	-0.88	-1.19	0.78

+

Residual error

-0.87	-0.43	1.53	-1.1
1.03	0.47	-1.27	0.8
-0.17	-0.03	-0.27	0.3
-2.07	-0.43	-0.57	-1.43
2.23	0.47	1.43	1.17
-0.17	-0.03	-0.87	0.27
-2.03	-0.53	1.5	-0.93
-0.53	0.67	-1.7	0.57
2.57	-0.13	0.2	0.37
-0.43	-0.93	0.77	-0.8
-0.03	0.97	-2.43	0.4
0.47	-0.03	1.67	0.4

Figure 12.7 Factor diagram of antimony data example.

243

To find the values to place in the antimony portion of the factor diagram (i.e., the effect of Treatment 1 or the first independent variable), we must subtract the grand average (19.55) from the averages for each level of the antimony factor. To make hand calculations easier, these values have been rounded. There are four levels of antimony (0, 3, 5, and 10), and the means for these levels, are 20.17, 20.41, 20.62, and 17.02, respectively. If you add together the treatment effects across each row in the Antimony portion of the diagram, the sum will be 0 (as some treatment effects will be above the mean and some will be below the mean, thus canceling each other out). Here, within this particular example, things do not work out in this way simply because of the way data values have been rounded. However, the sum is very close to 0.

To find the values to place in the *Method* portion of the diagram (i.e., the effect of Treatment 2 or the second independent variable), we find the averages for each of the four methods (which are, 18.64, 20.45, 20.18, and 18.95, respectively) and then subtract the grand average from the average for each level of method.

With a two-way ANOVA, we are also concerned about how the independent variables interact to affect the response. An interaction is said to occur when the effect of one independent variable on the response depends on the level of the second independent variable. To find the values to place in the *Interaction* portion of the factor diagram, we must first add the grand average to the estimated effects of our two independent variables (i.e., the values in the antimony and method portions of the factor diagram). This will give you what is called a *Partial Fit*, as you can see in Figure 12.8. You then subtract the Partial Fit from the cell averages for each of the different factor combinations—Antimony 0 and Method 1, Antimony 0 and Method 2, etc. to get the interaction effects.

Partial fit:

19.55	19.55	19.55	19.55	+	0.62	0.86	1.07	-2.53	+	-0.91	-0.91	-0.91	-0.91	=	19.26	19.5	19.71	16.11
19.55	19.55	19.55	19.55		0.62	0.86	1.07	-2.53		0.9	0.9	0.9	0.9		21.07	21.31	21.52	17.92
19.55	19.55	19.55	19.55		0.62	0.86	1.07	-2.53		0.63	0.63	0.63	0.63		20.8	21.04	21.25	17.65
19.55	19.55	19.55	19.55		0.62	0.86	1.07	-2.53		-0.6	-0.6	-0.6	-0.6		19.57	20.81	20.02	16.42

Cell averages (or factor combinations):

		Antimony			
		0	3	5	10
Method	1	18.47	19.03	20.77	16.3
	2	22.07	20.43	21.47	17.83
	3	20.33	22.23	21.4	16.73
	4	19.83	19.93	18.83	17.2

Interaction effects:

Cell averages					Partial fit					Interaction			
18.47	19.03	20.77	16.3		19.26	19.5	19.71	16.11	=	-0.79	-0.47	1.06	0.19
22.07	20.43	21.47	17.83	−	21.07	21.31	21.52	17.92		1	-0.88	-0.05	-0.09
20.33	22.23	21.4	16.73		20.8	21.04	21.25	17.65		-0.47	1.19	0.15	-0.92
19.83	19.93	18.83	17.2		19.57	20.81	20.02	16.42		0.26	-0.88	-1.19	0.78

Figure 12.8   Obtaining the interaction term in the antimony factor diagram.

Finally, to find the values of the *Residual Errors*, we simply take each observation and subtract the cell average from it (i.e., subtract the average of the particular factor combination that case happens to fall into from the value of the observation for that case). For example, our first observation is in antimony level 0 and method level 1, and it has an observed strength of 17.6. The cell average (or factor combination average) for that value is 18.47. If we take 17.6–18.47, we get −0.87.

Degrees of Freedom

To construct our ANOVA table, we need to determine degrees of freedom (df), sums of squares, mean squares, and our F statistics. We will share the ANOVA table shortly, but first, it is important to understand how to find degrees of freedom in a two-way design.

Remember that each part or section of our factor diagram has a particular value for degrees of freedom.

- For the *Grand Average*, the df is always equal to 1.
- For each treatment or independent variable, the df is equal to the number of levels of the variable minus one. Here, we know we have four levels of both *Antimony* and *Method*. Thus, we have $4 - 1 = 3$ df for both of our independent variables.
- To find the *Interaction* df, we simply multiply the dfs for each independent variable. Here, this would give us $3 \times 3 = 9$.
- For the *Residual Error*, we find the df by taking the number of factor combinations and we multiply this by the number of observations per cell (or factor combination) minus 1. Here, we have $16 \times (3 - 1) = 16 \times 2 = 32$. If there is an unequal number of cases per factor combination, you can also find the df error by taking N (or the total sample size)—number of factor combinations. Here, this would give us $48 - 16 = 32$.

Formula Approach and the ANOVA table

In the standard approach for one variable, we accounted for variation by means of the sum of squared deviations, or SS in the output table, from two sources:

- group or treatment effect (e.g., type of fat)
- residual error

In two-way ANOVA, we split the variation into four components:

- variable 1 effect (percent antimony)
- variable 2 effect (cooling method)
- interaction between variables 1 and 2
- residual error

Let us look at a typical software output, shown in Table 12.6. Keep in mind that as with the one-way ANOVA example described earlier in this chapter, the values within the SS column below were obtained by squaring values within each portion of the factor diagram and then adding together the squared values. For example, to find the grand average, we squared 19.55 a total of 48 times and then added all of those squared values together to arrive at a value of approximately 18,353.54. Although we include the grand average in the

TABLE 12.6 Two-way ANOVA table, software output

Source of Variability	SS	Df	MS	F	p-Value
Grand average	18,353.54	1			
Antimony	104.19	3	34.73	20.12	<0.001
Method	28.63	3	9.54	5.53	0.004
Interaction	25.13	9	2.79	1.62	0.152
Residual error	55.25	32	1.73		
Total	18,566.74	48			

following table, some statistical programs may omit this, and this is okay; it is mostly the effects of our two treatments and our interaction that we are especially interested in.

As Table 12.6 indicates, now there are three F-tests—one for each variable and one for the interaction between the variables. The first null hypothesis is that the percent of antimony does not affect soldering strength. The second null hypothesis is that the cooling method does not affect soldering strength. The third null hypothesis is that there is no interaction between percent of antimony and cooling method on soldering strength. The first two F-tests give very small p-values so we would probably reject both nulls and conclude that both variables have an effect. The third F-test leads to a larger p-value, however, and this suggests the absence of a significant interaction between the variables.

The table also shows a useful breakdown of the variability in solder strength. It shows the total solder strength value of 18,566.74, as well as how much of the strength is due to different levels of antimony, different cooling methods, the interaction between levels of antimony and cooling methods, and random error.

12.6 FACTORIAL DESIGN

The type of experiment we have been reviewing—an analysis of variance with multiple independent variables whose levels are set by the researcher—is called a *factorial design*. To see the advantages of factorial design, let us look at a hypothetical situation and consider other ways a study might be conducted. The hypothetical situation is that there are two types of subject, A and B and two treatments, 1 and 2. A factorial design produces a table like Table 12.7 of mean outcomes for the four factor combinations.

TABLE 12.7 Hypothetical outcomes of factorial design

	Treatment 1	Treatment 2
Subject type A	10	20
Subject type B	15	5

Clearly, there is a very strong interaction between treatment and subject type, meaning that the effect of one independent variable on the response depends on the level of the other independent variable. The outcomes differ markedly for subject types A and B. Now, consider other possible designs and how they would turn out.

Question 12.2

(1) One investigator recruits only subject type A. What will her investigation conclude about the effect of treatments 1 and 2? (2) A second investigator recruits only subject type B. What will her investigation conclude about the effect of treatments 1 and 2? (3) A third investigator recruits both subject types but keeps no record about which subject types get treatment 1 and which gets treatment 2. She has only the overall results. What will her investigation conclude about the effect of treatments 1 and 2?

The factorial design allows us to understand how effects differ for different groups and avoid the errors described earlier.

Stratification

In an experiment, the researcher can often control the levels of all the variables. In survey sampling, a researcher may not be able to alter a value for a subject but he or she may be able to choose subjects with appropriate values. Imagine a survey to be done in Vermont, for which it is believed race or ethnicity may be an important variable. We cannot assign a race or ethnicity to a subject. The demographics of Vermont are such that, in a general sample, it is apt to have many Whites but very few Blacks, native Americans, or Hispanics.

To overcome that problem, we can draw four separate samples from the four ethnic groups instead of from the population as a whole. In that case, the subpopulations are called *strata*, and the process is called *stratification*. It is limited in that it may not be practical to stratify on all variables at once if there are many of them. Fortunately, we can often pick and choose which variables we wish to stratify. Gender might also be a variable in the same survey, but the numbers of males and females in Vermont are balanced closely enough that stratification would not be needed. Stratification is needed, in other words, when the groups we need to sample make up a small part of the general population.

The strata then become variables in the regression. In the Vermont case, we might have a single categorical variable "stratum" that can take on the values, "White, Black, Native American, Hispanic." If these four categories are not provided in the form of binary variables, many statistics programs will convert them to that form. This example indicates how this is done, where $0 =$ no and $1 =$ yes.

White (0,1)
Black (0,1)
Native American (0,1)
Hispanic (0,1)
Caution: Be sure to read the note about multicollinearity in Chapter 13.

Blocking

Stratification is typically applied to observational or survey data. In designed experiments, the equivalent concept is known as *blocking*.

Suppose we want to study the impact of different levels of light and fertilizer on tomato plants. We may feel that this could vary with the breed of tomato, so we include the four most common varieties grown in our area. We cannot assign a breed to an existing tomato plant, but we can choose tomatoes of whatever breed we want. Although the idea is similar to stratification, there are some differences in practice. In stratified random sampling, every subject must belong to a stratum, and we take random samples from within each stratum.

Inference addresses generalization to the population. In experiments, there may be no attempt to include every possible value of the blocking variable, to take a random sample for each value of a blocking variable, or to use statistical inference to make generalizations about a population.

Often, the blocking variable, or the variable on which we stratify, is not one of the variables of interest in an experiment, but rather a "nuisance" variable. Such a nuisance variable is one that we can measure and include in the study but whose effect we are not interested in studying.

For the tomatoes, we may be primarily interested in the effect of light and fertilizer, but we fear that breed might also have an effect. By including breed in the design, we are able to account for some of the variability in our results. This allows us to make more precise estimates of the effects of the treatments in which we are interested. Sometimes, people say that "blocking reduces variability," but it is more accurate to say that it reduces unaccounted-for variability.

Analysis for a randomized block design is the same as that for any factorial design. However, we usually report only the results for the variables of interest and ignore the effects of the blocks.

Resampling Approach

A resampling approach to a two-way ANOVA is complex, both from the standpoint of specifying an appropriate procedure and from the standpoint of programming it. For this reason, in this chapter, we chose instead to focus on an extension of the standard formula approach used by most software programs.

12.7 EXERCISES

1. Use the antimony data discussed in this week's lesson. You can find the data at the book web site as a plain text file and as an Excel and CSV spreadsheet. Run a two-way ANOVA including interaction and use this model for all the following questions even if you decide the interaction term is expendable. Additional procedures may also be needed to fully answer the following questions.

   Important: The antimony data are available in two formats:

   - Standard statistical/database format for data, where each row is an observation.
   - Table form, in which each column is a different cooling method, each row is a specific level of antimony, and the values in the table are the measured strengths. There are three observations for each combination of antimony and cooling method.

   Excel needs to use the table format. StatCrunch needs to use the database format. The worksheet tabs indicate the format. The data are also available in non-Excel format (text and CSV).

   (a) Does the level of antimony affect joint strength?

   (b) Does the type of cooling affect joint strength?

   (c) Does there seem to be a problem with interaction here?

   (d) Explain in words what interaction would mean in terms of this study and these variables. Give an example.

   (e) Do the assumptions for ANOVA seem to be met here? Why or why not?

2. Web page load time depends on a number of variables, but internet service providers (ISP's) have control only over factors at the server end. A large ISP tests load time for the same page using three different server configurations at different times, with the following results:

Config. A	Config. B	Config. C
1.049	1.979	0.948
1.029	1.992	0.856
1.349	2.323	1.004
1.171	2.056	0.386
2.071	2.308	0.308
0.366	3.558	1.589
1.201	2.206	0.274
1.434	2.291	0.767
0.836	1.962	0.978
3.201	1.112	0.371
0.415	2.368	0.298
1.934	1.178	0.445

(a) Calculate the average load time at the three server configurations

(b) Decompose the data into factor tables as was done in Figure 12.3.

(c) Test the hypothesis that the three servers do not differ with respect to load time.

3. The web server experiment is repeated, this time with two different databases to power the web page being tested—MYSQL (top block) and PostgreSQL (bottom block)

Config. A	Config. B	Config. C
MYSQL		
1.021	2.044	0.909
1.028	2.008	0.938
1.407	2.251	1.097
1.18	1.957	0.408
2.015	2.231	0.26
0.307	3.65	1.58
PostgreSQL		
1.305	1.921	0.414
1.556	1.954	1.038
1.025	1.571	1.191
3.347	0.816	0.579
0.656	1.982	0.511
2.234	0.952	0.592

(a) What is the grand average?

(b) What is the configuration effect?

    (c) What is the database effect?

    (d) What is the interaction effect?

    (e) Using software, determine the F-value for each of b, c, and d.

    (f) Put the analysis all together in a few sentences of interpretation that would make sense to a nonstatistician.

Answers to Questions

12.1 Using one big batch that is properly mixed ensures that the dough used for each replication is the same. Mixing separate batches could introduce some variation in the composition of the dough. The main constraint on using a big batch is the ability of the cook and his equipment to handle the amount of dough required.

12.2 The first investigator will conclude that Treatment 2 has a larger effect, the second investigator will conclude that Treatment 1 has a larger effect, and the third investigator will have lots of conflicting results, with, on balance, no difference between Treatments 1 and 2.

13

MULTIPLE REGRESSION

In the previous chapter on ANOVA, we had an outcome variable, joint strength, and one or more input variables, antimony, and cooling method. We were concerned with the question, "*Is there* a statistically significant effect of either input variable on the outcome variable?"

We might also be interested in this question, "*How much* does the outcome change as you change the input variables?"

After completing this chapter, you should be able to

- distinguish between regression for explanation and regression for prediction,
- see how regression models the relationship between a supposed outcome variable, and supposed predictor variables, but that the mathematics of regression does not *prove* any causation,
- use visual exploration data to determine how appropriate a linear model would be (and for what range of the data),
- perform a multiple linear regression,
- interpret coefficients and their p-values,
- use R^2 to measure of how well the regression model fits the data,
- use root mean squared error (RMSE) as a measure of prediction error,
- explain the procedure of partitioning the data into a training sample (to fit the model) and a holdout sample (to see how well the model predicts new data),
- use resampling to establish confidence intervals for regression coefficients,
- use regression for the purpose of predicting unknown values, using the Tayko data,
- describe how binary and categorical variables are used in regression, how the issue of multicollinearity arises, why it may be important, and what to do about it.

Introductory Statistics and Analytics: A Resampling Perspective, First Edition. Peter C. Bruce.
© 2015 John Wiley & Sons, Inc. Published 2015 by John Wiley & Sons, Inc.

The "how much" question typically arises in two contexts.

(1) We want to understand a phenomenon and explain how variables relate to one another, for example, how does global warming depend on carbon dioxide emissions?

(2) We want to predict individual values, for example, how much will a consumer spend, if you send her a catalog?

Prediction or Explanation

Regression is a ubiquitous procedure—it is used in a wide variety of fields. Most uses fall into one of the two categories:

1. *Explaining relationships. Researchers* want to understand whether x is related to y. For example, is race a factor in criminal sentencing? How do sex and other factors, affect earnings?
2. *Predictive modeling. Data scientists* and business analysts, for example, want to predict how much a customer will spend?

13.1   REGRESSION AS EXPLANATION

For answers, we will turn to regression, a technique that was introduced earlier for one variable and which we will now extend to multiple input variables. The term *"input variable"* is used here; elsewhere, the term *"independent variable"* is used. Essentially, we are attempting to predict or explain the behavior of a variable of interest—the outcome—in terms of the levels of other variables—inputs. In different fields, such as biostatistics, data mining, and machine learning, these variables have different names (Figure 13.1).

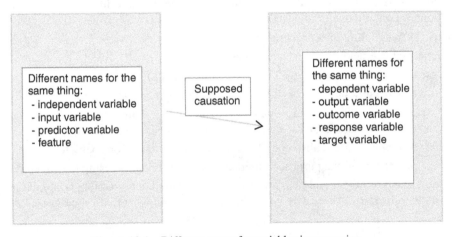

Figure 13.1   Different terms for variables in regression.

Here is a summary.

In these problems, we have measurement data on an outcome variable—a single dependent variable of interest. We wish to model how that variable depends on other variables—input or independent variables. We also wish to measure how reliable our model is—how much it might differ if we were to select a different dataset. We may also wish to test whether an apparent relationship between input variables and an outcome variable could be due to chance. We will first review simple linear regression and address such problems via a resampling technique. Then, we will discuss the process of going from the single independent variable you already know about, to multiple independent variables.

 The diagram shown earlier talks of "causation" and the terminology refers to one outcome variable "depending" on other variables. The directional nature of this relationship is a product of our belief, presumably on the basis of theory or knowledge, but regression does not prove it. The mathematics of regression merely describes a relationship; it does not prove a direction of causation. So the logical train of thought is thus the following: (i) We have a theory that y depends on a set of x-variables. (ii) Regression analysis may confirm that there is a relationship, and it may also describe the strength of that relationship. (iii) If so, we take this as evidence that our theory is correct. However, you can see that there is no guarantee that the theory that y depends on x is correct. The direction of the relationship could be the reverse. Or both x and y could depend on some third variable.

13.2   SIMPLE LINEAR REGRESSION—EXPLORE THE DATA FIRST

We ended our introduction to ANOVA with a discussion of data on the strength of soldered joints. Let us revisit simple linear regression (regression with one input variable) to estimate the effect that antimony quantity has on strength. The issue is whether adding more antimony, to lower costs, will reduce joint strength. Although the data and the problem are real, we will add some fictionalized details to illustrate how normal research includes a multiplicity of approaches. We will encounter issues that rarely show up in a final published report but are essential parts of the research process.

Antimony is Negatively Correlated with Strength

One of the engineers at the factory using the solder has objected to using antimony all along. He provides this printout, which shows accurate values.

```
Correlation of antimony and strength = -0.570, p-value = 0.000
```

The engineer says that the moderately strong negative correlation shows that strength decreases with antimony content, while the very low p-value leaves little doubt that the effect is real.

Is There a Linear Relationship?

We saw earlier that the correlation coefficient and linear regression do a good job only when the relationship being studied is linear. Let us look at the data now to see if, in fact, the relationship between antimony and strength is linear. Figure 13.2 is a scatterplot of the two variables.

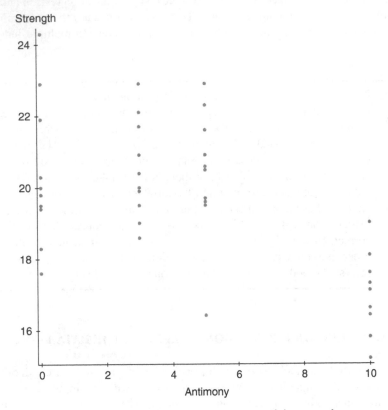

Figure 13.2   Scatterplot of antimony versus joint strength.

It appears that the relationship is *not* linear. The negative correlation between joint strength and antimony is true only as antimony is increased to the highest level.

Another view is afforded by the set of boxplots in Figure 13.3.

These boxplots provide further confirmation. As antimony increases from 0% to 3% to 5%, joint strength changes a little. If anything, it increases slightly. At 10%, joint strength drops off precipitously. Thus, there is no simple linear relationship between antimony and joint strength.

At this point, a group of managers meet to discuss the results. The graphs shown earlier suggest that antimony's negative impact on joint strength may be limited to the 10% level. The managers decide that the 10% level is now off the table. They ask the research team to look only at the data for the other antimony levels. Overall, it looks like good news. If antimony seems to have little impact on joint strength at low levels, then we can use some and cut costs. In that case, it looks like we can get away with as much as 5%. The next step is to examine the relationship between antimony and joint strength in the limited range

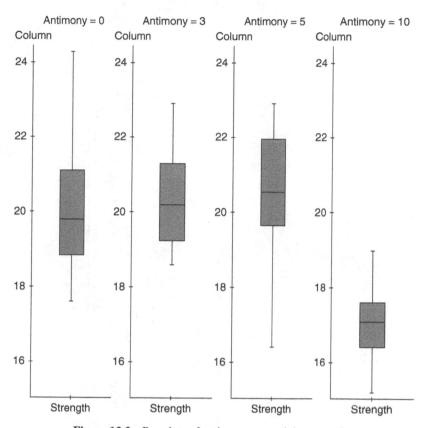

Figure 13.3   Boxplots of antimony versus joint strength.

between 0% and 5%. We can use linear regression for this; the Excel output for a model with antimony concentration as an input variable and joint strength as an outcome variable is shown in Figure 13.4.

The regression equation is

```
S = 20.2 + 0.087 * A
```

Try It Yourself 13.1

Interpret the equation shown above in words.

The slope, 0.087, is now positive rather than negative, meaning that adding antinomy in this range seems to increase strength. The p-value of 0.528 tells us, though, that the positive slope could well be due to chance. In any case, we have no evidence that antimony *reduces* strength over this range.

We reproduced the figure of an ANOVA table shown earlier to show that regression and ANOVA are intimately related. The table was created by the software that ran this regression. Note that the p-value for the t-test on the antimony coefficient and the p-value

SUMMARY OUTPUT

Regression statistics	
Multiple R	0.108651
R Square	0.011805
Adjusted R Square	–0.017259
Standard error	1.692685
Observations	36

ANOVA

	df	SS	MS	F	Significance F
Regression	1	1.16375	1.16375	0.406169	0.52818877
Residual	34	97.41625	2.865184		
Total	35	98.58			

	Coefficients	Standard error	t Stat	P-value	Lower 95%	Upper 95%	Lower 95.0%	Upper 95.0%
Intercept	20.16667	0.462203556	43.63157	1.94E-31	19.227356	21.10598	19.227356	21.105977
Antimony	0.0875	0.137294915	0.637314	0.528189	–0.19151684	0.366517	–0.191517	0.3665168

Figure 13.4 Excel regression output.

for the F-test have the same value of 0.528. The F-test is for the null hypothesis that joint strength does not differ from one antimony group to another. The t-test is for the null hypothesis that the regression slope of strength as a function of antimony level is zero. In this case, the two hypotheses are equivalent.

Back at the soldering plant, the research team has reported these results to management. Because the first three levels of antimony seem to give about the same joint strength, the resulting action is to adopt 5% antimony solder for current use. This maximally reduces cost at no loss of joint strength. Management also forwards the research report to the solder manufacturer, suggesting that it explores additional amounts of antimony between 5% and 10% in the interest of lowering costs even more.

13.3   MORE INDEPENDENT VARIABLES

The years 2002–2005 marked a housing bubble around the world, particularly in the United States. In 2006, prices started declining. In 2007, the decline accelerated, culminating in 2008 in a financial panic, massive intervention in financial markets by the US Federal Reserve, and the start of a protracted and severe recession.

In retrospect, the growth of the bubble is clear. At the time, however, key actors failed to see it or ignored the signs. Before the bubble collapsed, many people concluded that their house was fairly priced if all other similar houses nearby were similarly priced.

Is it possible to establish an objective standard for home valuation so that lenders and real estate brokers have a reference point more solid than the other houses comprising the bubble?

One possibility is to consider the underlying determinants of home value and then establish a relationship between those variables and home prices. Until now, we have considered a single independent predictor variable and a single dependent outcome variable. Now, we must incorporate multiple independent variables.

Multiple Linear Regression

It is unlikely that a phenomenon of interest—housing prices, in this case—can be explained by a single variable. Usually, multiple factors are at work, and we can model those in a multiple linear regression of the form

$$y = a + b_1 x_1 + b_2 x_2 + \cdots + \varepsilon$$

In words, this means that the dependent variable y is a function of a constant, a, plus a series of independent variables x_1, x_2, and so on, times their respective coefficients b_1, b_2, and so on, plus an error term ε.

When we had just one independent variable, we saw that the chosen regression line minimized the sum of the squared deviations between the actual y-values and the predicted y-values that are points on the line. With multiple variables, the mathematics is more complex because we are no longer dealing with a line. However, the idea remains the same, which is to minimize squared deviations between predicted and actual values.

Notation

The dependent, that is, outcome, variable is typically denoted by y.

Predicted values of y are typically indicated by the symbol $\hat{y}$ or "y-hat."

Independent variables in the abstract are usually referred to as "x-variables"—$x_1, x_2, \ldots, x_i$.

The coefficients belonging to these x-variables are typically denoted by letters—$b_1, b_2, \ldots, b_i$ or by the Greek letter beta—$\beta_1, \beta_2, \ldots, \beta_i$.

The error term, also called the *residual*, is usually denoted by the letter "e" or the Greek letter epsilon—ε.

Let us consider a simplified example—the Boston Housing data from the 1970 census. The complete data are available at the UCI Machine Learning Repository: Housing Data Set—http://archive.ics.uci.edu/ml/datasets/Housing. A reduced set of variables is used here. The outcome variable is the average price of homes by census tract, which is a neighborhood defined by the US Census Bureau. A tract consists of several thousand people and is relatively homogeneous with respect to demographic characteristics. The independent variables are factors that are believed to have an effect on home value. For simplicity, this presentation uses only a subset of the records and a subset of the variables originally associated with the data set (Table 13.1).

TABLE 13.1   Boston Housing Data Variables

Independent (predictor) variables	
CRIM	Crime rate per 1000 residents
RM	Average number of rooms per dwelling
Dependent (outcome) variable	
MEDV	Median home value in $1000s

As a first step, it would be useful to understand how these variables are distributed. We can do this with a set of separate boxplots, shown in Figure 13.5. Note that the variables do not share the same y-axis.

From the first boxplot, CRIM, the crime rate, we can see that most neighborhoods have very low crime rates with little variation among them and that there are some neighborhoods with much higher rates. You should take a moment to review the other boxplots to get a sense of how the variables they represent are distributed.

Next, it would be useful to learn the extent of correlation among the variables, especially the correlation between each independent variable and the dependent variable, which is median home value. We can do this with the correlation matrix shown in Figure 13.6.

Definition: **Correlation matrix**

In a correlation matrix, the variable names constitute both the row and column headings. Each cell in the matrix indicates the correlation between its row and column variable.

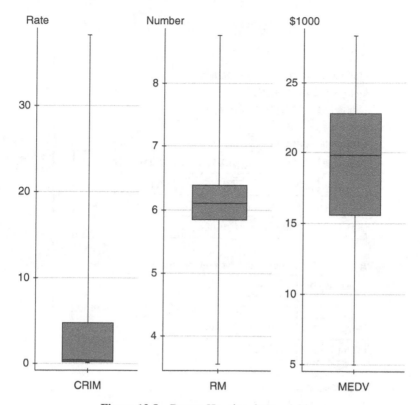

Figure 13.5   Boston Housing data variables.

	CRIM	RM	MEDV
CRIM	1		
RM	−0.11605	1	
MEDV	−0.60046	0.36591	1

Figure 13.6   Boston Housing data: correlation matrix.

For example, the correlation between RM, the number of rooms in a home, and MEDV, the median value of the home, is 0.36591, which is a low positive correlation. Also, the correlation between CRIM, the crime rate, and MEDV is −0.60046. As the crime rate decreases, the median value of the home generally increases. Along the diagonal is a series of 1s; the correlation of a variable with itself is 1. In this presentation, the cells above the diagonal are left blank as they are duplicative of the cells below the diagonal (the correlation of RM with CRIM is the same as the correlation of CRIM with RM).

Finally, using software, we can perform a multiple regression analysis on the data. We specify MEDV as the dependent—outcome—variable and CRIM and RM as the predictor—independent—variables. Figure 13.7 shows the output of the multiple regression analysis from StatCrunch.

Let us focus now on the coefficients and their interpretations. Ignore the other part of the output for now.

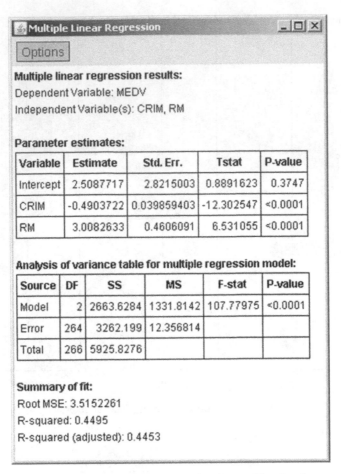

Figure 13.7   Multiple linear regression: Boston Housing data.

The CRIM coefficient is −0.4903722, and the RM coefficient is 3.0082633. The y-intercept is 2.5087717, but this is not a very meaningful number in this case.

Question 13.1

Why is the y-intercept not a meaningful number in this case?

We will round these coefficient values to four significant figures for brevity.
The regression equation is:

$$\text{MEDV}_r = -0.4904 * \text{CRIM} + 3.009 * \text{RM} + 2.509$$

We should be able to use this equation to predict the MEDV for a different set of housing tracts. We will do this shortly, in the context of model assessment.

13.4   MODEL ASSESSMENT AND INFERENCE

The next question is, "How good a model is this?" We will look at the following questions.

1. How well does it explain the data that went into building the model?
2. How well does it predict new data? (See the following section on inference via holdout sample.)
3. How reliable is the model as an explanation for the observed data, given the possible role of chance? (See the following on bootstrapping a regression.)
4. To what extent are the statistical assumptions underlying regression met?

R^2

R^2 is the measure of the extent of variation in y that is explained by the regression model. We already encountered it for the case of simple linear regression: R^2 is the square of the correlation coefficient ρ. An R^2 value of, for example, 0.75, means that the regression model explains 75% of the variation in y.

R^2 is a biased estimator. Its value in a sample will always be higher than the value you would get if you could perform a regression on the entire population. You can see this most easily in a small sample for a regression with multiple predictor variables. Consider the simplest case of all—two observations and two variables. The regression line will fit perfectly, and R^2 will be 1.0, but such a model tells you next to nothing about the data.

Many software programs produce an adjusted R^2 that corrects for both the size of the sample and the number of predictor variables. The more predictor variables you have, the larger is the sample required to get a useful regression estimate.

Inference for Regression—Holdout Sample

Another measure of the value of a regression model is how well it predicts new data. The only way we can measure this is to apply the regression model to some additional data where the outcome variable is known, so we can see how well the predictions did.

In this case, we have such a sample—a special one called a "*holdout*" sample. Boston-HousingSmall_Holdout.xls was created before we started this problem by randomly selecting one-third of the records in the original dataset and not using them in the analysis we have just gone through.

Definition: Training and holdout samples

Records used in building regression models can be randomly partitioned into training data and holdout data.

Training data are the records used to fit the model. Usually, they constitute 50–70% of the data.

Holdout data are the records to which the model is applied, to see how well it does. The model does not "see" the holdout data until after the model has been fit to the training data.

As this holdout sample is from the same source as the original sample, and as it contains known values for MEDV, we can use it to answer the question, "How well did the model do in predicting MEDV for this new data?" We can compare the predicted values with the actual values, knowing that these new data were not used in building the model. Here are the first few rows of the holdout sample, showing the predicted value in column C, the actual value in column D, the residual or error in column E and the residual squared in column F (see Figure 13.8).

C2			f_x	=A2*-0.4903722+B2*3.0082633+2.508717				
	A	B	C	D	E	F	G	H	I
1	CRIM	RM	Pred. MEDV	MEDV	Residual	Residual Sq.		RMSE	4.2712152
2	0.2498	5.857	20.0056202	13.3	6.7056202	44.9653419			
3	0.06466	6.345	21.5644402	22.5	-0.9355598	0.875272192			
4	0.05059	6.389	21.7037033	23.9	-2.1962967	4.82371922			
5	0.1676	6.426	21.7576306	23.8	-2.0423694	4.171272827			
6	0.25356	5.705	19.5465204	16.2	3.3465204	11.19919846			
7	5.73116	7.061	20.9396826	25	-4.0603374	16.48633961			
8	5.44114	6.655	19.8605255	15.2	4.6605255	21.72049765			
9	0.32543	6.431	21.6952765	18	3.6952765	13.6550681			
10	0.04527	6.12	20.8970892	20.6	0.2970892	0.08826202			
11	10.0623	6.833	18.1299079	14.1	4.0299079	16.24015801			
12	0.15936	6.211	21.1148946	24.7	-3.5851054	12.85298042			
13	4.55587	3.561	10.9870706	27.5	-16.512929	272.6768368			
14	0.16902	5.986	20.4332984	21.4	-0.9667016	0.934511975			
15	1.15172	5.701	19.0940546	13.1	5.9940546	35.92869059			
16	0.32264	5.942	20.2256038	17.4	2.8256038	7.984037072			
17	4.81213	6.701	20.3073546	16.4	3.9073546	15.26741996			

Figure 13.8 Predictions shown in column C from applying regression model developed from BostonHousingSmall.xls to the data in BostonHousingSmall_Holdout.xls.

An overall measure of error is "root mean squared error" or RMSE, using the residuals (predicted vs actual values). It is calculated by squaring the residuals, taking the average, and then finding the square root. You can think of this as the "typical" error in each predicted value.

Definition: **Root mean squared error**

$$\sqrt{\frac{\sum_1^n (\hat{y}_i - y_i)^2}{n}}$$

In this model, the RMSE is calculated by finding the difference between the predicted MEDV and the actual MEDV—the residuals in column E. The residuals are squared and stored in column F. The RMSE in cell I1 is calculated by the formula = SQRT(AVERAGE(F2 : F134)).

The calculated RMSE for the values we predicted using the main regression equation is 4.271 or $4271. We obtain this value by multiplying (4.271 * $1000) as MEDV is denominated in $1000.

Try It Yourself 13.2

In the following data, cholesterol levels are predicted based on the number of miles walked per day

Miles	Chol
1.5	193
0.5	225
3	181
2.5	164
5	140
3.5	211
4.5	158
2	178

The scatterplot is shown below:

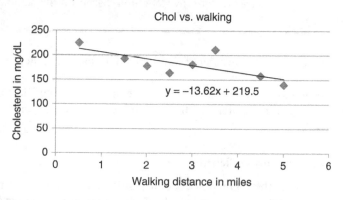

The regression equation is:

```
cholesterol = -13.628*miles + 219.58
```

Using the previous Boston Housing calculations as an example, calculate the following (in Excel):

1. The predicted cholesterol for each value of Miles
2. The residuals and squared residuals
3. The RMSE

Model Assessment via Holdout Samples

Using a holdout sample to assess the model has two great virtues.

1. *Conceptual simplicity*. The idea of assessing your model on fresh data that was not used initially is intuitively appealing. It is also easy to explain to colleagues who

may be involved in your project but who may not have a great deal of statistical background.

2. *Clean inference.* Complete separation of the data used to build the model from the data used to assess it gives a less biased evaluation of the model. In traditional inference, including both the resampling and formula methods we have discussed to this point, we assess the model using the same data that were used to build it. Thus, we do not get an independent view.

There are also two drawbacks.

1. Holdout samples are costly in their use of data. If data are scarce, most analysts want to use all of it to build as accurate a model as possible. This leaves nothing to use for a holdout sample.

2. Holdout samples are good at assessing how good the model's predictions are, but they do not produce the traditional measures of the model itself—confidence intervals for coefficients and p-values. Those measures are, by definition, based on the sample used to fit the model.

Holdout samples are used on occasion in scientific *research*, but that community will likely want to see confidence intervals and *p*-values.

In data science, the use of holdout samples is the rule. In predictive modeling, data are typically plentiful and the main goal is accurate prediction. Holdout samples let you measure that directly.

Confidence Intervals for Regression Coefficients

Let us get back to the question of the regression coefficients. If you are interested not just in prediction but also in explanation, you would likely be interested in the coefficients and how reliable they are. If we took a different sample of housing tracts and performed a new regression, would we see completely different coefficients or would they be similar?

The answer to that question, familiar by now, is given by drawing repeated bootstrap samples from our original sample, and repeating the regression over and over, recording the coefficients each time.

Bootstrapping a Regression

Here is the procedure applied to the Boston Housing data.

1. Place N slips of paper in a box, where $N =$ the original sample size. $N = 267$ for BostonHousingSmall.xls. On each slip of paper, write down the values for a single census tract. Here is how the slip of paper might look for one tract.

CRIM	RM	MEDV
8.98296	6.212	17.8

2. Shuffle the papers in the box, draw a slip, record its values, and replace.

3. Repeat step two N times.

4. Using this resampled data, perform a regression with MEDV as dependent variable and the others as the independent variables. Record the coefficients.

5. Repeat steps 2 through 4 1000 times.

You will end up with distributions of 1000 values for each coefficient.

The bootstrapped values for the coefficients shown above and the resulting confidence intervals are found in bostonhousingsmall_rsxl.xls. This worksheet also contains the Resampling Stats setup for this problem. The specific procedure for running this model can be found in the textbook supplements.

Definition: **Bootstrap confidence interval—regression coefficient**

A 90% bootstrap confidence interval for a regression coefficient is the interval that encloses 90% of the bootstrap regression for that coefficient, cutting off 5% at either end. A 95% interval would cut off 2.5% at either end, and a 99% interval would cut off 0.5% at either end.

Software note: Of the software programs used extensively in this book and the supplements, only R and Resampling Stats can easily "enclose" a regression inside a resampling procedure. Most comprehensive statistics programs now include an option for bootstrapping. The more facilities the program has for user-based scripting, the more flexible the bootstrap option will be.

Inference for Regression—Hypothesis Tests

We have just finished looking at confidence intervals. Now, let us look at hypothesis testing for regressions. Specifically, we will be testing the null hypothesis that the regression coefficients are equal to zero, that is, that there is no relationship other than what chance might produce.

Typical regression output will include *p*-values for coefficients. Let us look again at the Boston Housing regression output, which we have seen earlier in Figure 13.7 and is repeated below as Figure 13.9.

Consider the row for "RM."

- 3.008 is the estimated regression coefficient, which we covered earlier.
- 0.4606 is the standard error of the estimated coefficient RM. You can think of this as equivalent to the standard deviation of the bootstrapped coefficients, which we have produced earlier.
- 6.531 is the value of the *t*-statistic.
- <0.0001 is the *p*-value.

You can think of the *p*-value in two ways.

1. Imagine repeated regressions in which you shuffle the MEDV variable prior to performing the regression. Shuffling destroys any relationship between RM and MEDV,

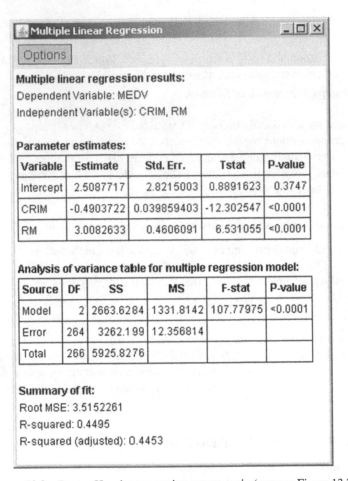

Figure 13.9 Boston Housing regression output again (same as Figure 13.7).

so any resulting coefficient will reflect chance only. How likely are you to see an estimated coefficient as high as 3.008? This is the estimated p-value. The very low p-value, therefore, tells us that the observed value of 3.008 is unlikely to have come about by chance. John Elder of the data mining firm Elder Research, uses this "target shuffling" approach to assess whether patterns discovered during data mining are real, or might be the product of chance.

2. Consider the bootstrap confidence interval for RM, which we have derived earlier. If zero lies outside this confidence interval, then the estimated coefficient for RM of 3.008 is statistically significant, which is equivalent to a low p-value.

The t-statistic is the formula equivalent #1. Its derivation is beyond the scope of this course but earlier in this book, we discussed the relationship of the t-distribution to the permutation distribution. The logic is similar here.

The F-statistic reported in the Model row is the same F-statistic that we discussed earlier in the context of ANOVA. It indicates whether there is an overall statistical significance that considers all the variables.

13.5 ASSUMPTIONS

Multiple linear regression models work best when certain requirements, often called *assumptions*, are met. The nature of these requirements, as well as methods for checking them, are listed below.

 This section will be important mainly to *researchers* producing studies that purport to explain or illuminate. Journal reviewers and regulatory authorities like explanations that are complete, robust, and long-lasting. If the requirements of regression models are not met, they will worry that something has been overlooked or that the results will not hold up over time.

Data scientists need not worry so much about these requirements—they will be interested mainly in whether the model is good at predicting.

Assumption 1: The observations are independent.

This is usually assured by design and can rarely be checked from the collected data. In a survey, we must have one of the two conditions. The first is a simple random sample. The second is a complex survey design in which the sources of lack of independence, such as two subjects belonging to the same stratum, can be modeled in the regression. In an experiment, we need random assignment of treatments to subjects. An observational study is a challenge because the researcher has little control over how the data were collected. Thus, the best that can be done is to investigate and fully disclose the data collection process. These issues apply equally to both traditional formula procedures for tests as well as to resampling.

Assumption 2: The relationship being investigated is, indeed, linear.

Assumption 3: The variance of y does not change as x changes.

Assumptions 2 and 3 are both tested by plotting residuals against predicted values. The resulting scatterplot should essentially be rectangular and random along the vertical axis, although some vertical stripes might be more dense than others. Figure 13.10 illustrates such a distribution produced by generating random numbers.

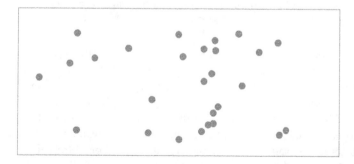

Figure 13.10 Random x and y coordinates, plotted.

The scatter diagram in Figure 13.11, which is neither rectangular nor random, indicates some basis for questioning whether assumptions 2 and 3 hold. All the residuals below a MEDV of 12 are negative.

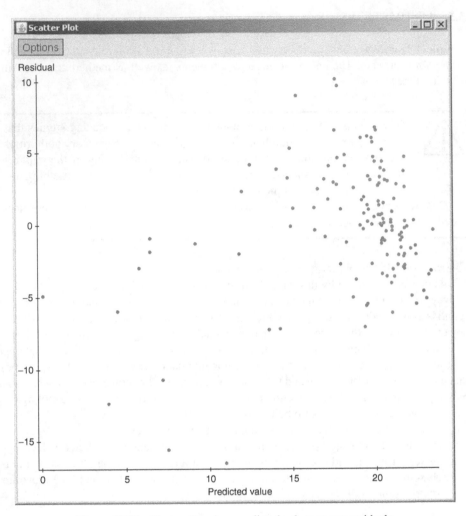

Figure 13.11   Boston Housing, predicted values versus residuals.

Assumption 4: For every value of x, the values of y are Normally distributed.

The fourth assumption, that the errors are Normally distributed at each level of x, is most important for traditional inference techniques, although these techniques still work well for mild departures from Normality. It applies equally to simple or multiple regression as well as to ANOVA. The usual check is a plot of the residuals. A Normal probability plot (a QQ plot in StatCrunch) is the most sensitive of commonly used residual plots, but it is also the hardest to interpret. We want the points to look like they are on a straight line. Figure 13.12 shows the residuals plot for the Boston Housing data.

We will not go into the details of how this plot is constructed because different software does it in different ways. Instead, we simply note that if the errors, or residuals, are distributed Normally, they will lie along the straight diagonal line. In checking for departure from Normality, we look for the graph drooping or rising at the left or right end, and we look for sudden jumps or gaps in the graph. In the plot shown earlier, we have evidence of a departure from Normality. Normal probability plots are very sensitive in detecting a

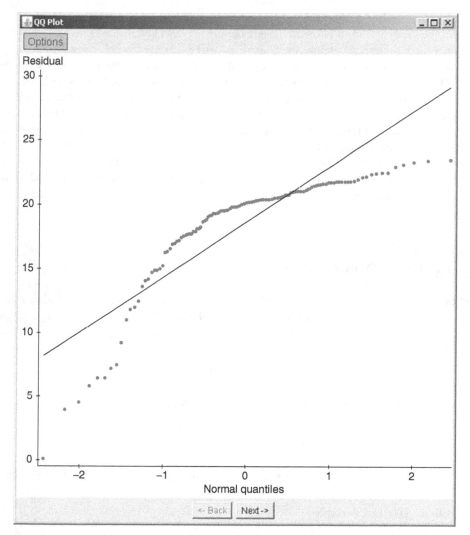

Figure 13.12   StatCrunch QQ plot for residuals (Normal probability plot).

problem. However, you need a lot of experience with these plots to be able to say what the problem might be.

Violation of Assumptions—Is the Model Useless?

If residuals are randomly and Normally distributed, this is evidence that the model has effectively accounted for all nonrandom sources of variation—a perfect model. Perfect models are hard to achieve. The worlds of natural phenomena and human behavior are complex and difficult to explain or predict completely. Usually, we must make do with models that fall short of 100% perfection—models where we lack strong evidence that the residuals are randomly and Normally distributed, or where we know, in fact, that they are not so distributed.

What happens if the assumptions are not met? Is the model useless? Not at all. Violation of assumptions detracts from the strength and believability of the model as an explanation of the relationship between the independent variables and the response variable. The greater the departure from Normality, for example, the less believable is the model, but it does not render the model valueless. Some knowledge is better than no knowledge.

In the housing data, for example, the residuals seem to violate assumptions at the extreme ends of the data. They are predominantly negative at predicted values below \$12,000 and above \$22,000. However, they are better behaved in the mid-range between these values. So, the model may be useful, after all, for nonextreme predicted values of MEDV. This is often the case with regression models—there may be a range over which the regression model is a good explanation of the data, and other range or ranges over which it is less good.

Moreover, when regression is used for prediction rather than explanation, and a holdout sample is used to assess performance, the importance of these assumptions diminishes. In that case, our main concern is whether the model does a good job in predicting unknown values.

13.6   INTERACTION, AGAIN

Let us now consider how we incorporate an interaction term into a regression model. Without interaction, a regression equation with two independent variables x_1 and x_2 and the random error term "ε," the Greek letter epsilon, looks like this.

$$y = a + b_1 x_1 + b_2 x_2 + \varepsilon$$

With an interaction term added, the equation becomes

$$y = a + b_1 x_1 + b_2 x_2 + b_3 x_1 x_2 + \varepsilon$$

In Excel, you create the derived variable $x_1 x_2$ by multiplying the two variables together.

Definition: **Derived variable**

A derived variable is a variable that is created from two or more other variables. A derived variable can be included in a regression model and treated like any other variable.

Let us look again at the Boston Housing data and create a new derived variable from CRIM, the crime rate, RM, and the number of rooms: CRIM*RM. The new regression output is shown in Figure 13.13.

This is the resulting regression equation, rounded to four significant figures.

$$MDEV_i = CRIM * 1.247 + RM * 4.680 - CRIM * RM * 0.2916 - 7.624$$

Compared to the initial regression model without the interaction term, the effect of CRIM is now positive. Although the value is small, a higher crime rate predicts a higher MEDV. RM is slightly larger (more rooms increase value). The interaction term is significant—it has a very low p-value—and its effect is negative.

Let us look at the effect of increasing RM by 1 unit.

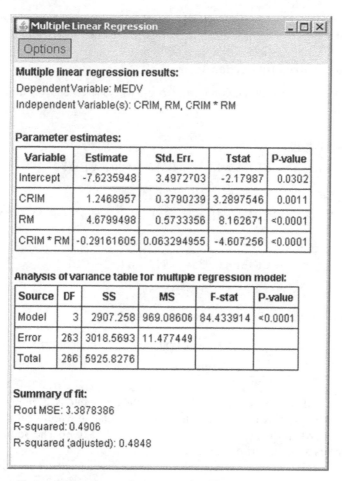

Figure 13.13   Regression output with CRIM*RM interaction.

Original Regression with No Interaction Term

For the original regression, when RM is increased by 1 unit, the response changes as follows.

Original Eq :

$$\text{MEDV}_r = \text{CRIM} * -0.4904 + \text{RM} * 3.008 + 2.509 \tag{13.1}$$

$$\text{MEDV}_{r+1} = \text{CRIM} * -0.4904 + (\text{RM} + 1) * 3.008 + 2.509 \tag{13.2}$$

$$= \text{CRIM} * -0.4904 + \text{RM} * 3.008 + 3.008 + 2.509 \tag{13.3}$$

$$= \text{CRIM} * -0.4904 + \text{RM} * 3.008 + 5.517 \tag{13.4}$$

The effect of increasing RM by 1 unit is:

$$\text{MEDV}_{r+1} - \text{MEDV}_r \text{ or Equation } 13.4 - \text{Equation } 13.1$$

$$= 3.008$$

So, if we hold the crime rate constant with no interaction term, then increasing the average number of rooms by one increases the average value of homes in the tract by $3008.

The Regression with an Interaction Term

For the regression with an interaction term, when RM is increased by 1 unit, the response changes as follows.

Regression Eq :

$$MDEV_i = CRIM * 1.247 + RM * 4.680 - CRIM$$
$$* RM * 0.2916 - 7.624 \tag{13.5}$$

$$MDEV_{i+1} = CRIM * 1.247 + (RM + 1) * 4.680 - CRIM * (RM + 1)$$
$$* 0.2916 - 7.624 \tag{13.6}$$

$$= CRIM * 0.9554 + RM * 4.680 - CRIM$$
$$* RM * 0.2916 - 2.944 \tag{13.7}$$

The effect of increasing RM by 1 unit is:

$$MEDV_{i+1} - MEDV_i \quad \text{or Equation 13.7–Equation 13.5}$$
$$= -0.2916 * CRIM + 4.68$$

For $CRIM = 3$, this works out to 3.805—when the crime rate is at three, adding a room adds $3805 to MEDV. This is more than the original equation without an interaction term yields. However, when $CRIM = 7$, this works out to only $2639. The value of additional rooms diminishes at higher crime rates.

Does Crime Pay?

Let us return to the coefficient for CRIM: 1.247. When the interaction term is included, it looks like higher crime rates produce higher home values, which makes no sense. But we must look at the whole picture, taking account of the interaction term.

Try It Yourself 13.3

What is the effect of increasing CRIM by 1 unit when $RM = 5$? What about when $RM = 7$? Do higher crime rates really increase home values?

13.7 REGRESSION FOR PREDICTION

So far we have been concerned with the situation in which the outcome variable is known, and we want to understand how it depends on various independent variables. Let us now consider the situation where the outcome variable is not known, and you want to predict it. Here is an example.

13.7.1   Example

Prediction task: You have access to lists of potential customers and a catalog that is relatively expensive to print. You would like to mail it only to those customers who are predicted to spend more than a certain amount.

How can this be done? If you do not have any known *y*-values—outcomes—you cannot create a regression model. The only solution is to build a model using data where the outcome variable *is* known and then apply it to the data where you need the prediction and where the outcome variable is unknown.

Tayko

Consider the hypothetical case of Tayko Software, a company that sells games and educational software. [(c) Statistics.com, used by permission.] It has recently put together a new catalog, which it is preparing to roll out in a direct mail campaign. In an effort to expand its customer base, it has joined a consortium of similar catalog firms. The consortium affords members the opportunity to mail catalogs to names drawn from a pooled list of customers totaling more than 20 million. Members supply their own customer lists to the pool and can borrow an equivalent number of names each quarter. A member can also apply predictive models to the entire consortium database to optimize its selection of names. Tayko has supplied 200,000 names from its own customer list. It is therefore entitled to mail to 200,000 names from the consortium per quarter. However, it has decided on a more limited effort and has budgeted funds for a mailing to 50,000 names. Although Tayko is a hypothetical company, the data in this case are from a real company that sells software through direct sales. The concept of a catalog consortium is also real, based on the Abacus Catalog Alliance.

The task is to build a regression model that will predict how much a customer will spend, based on a small sample. Tayko will then apply this model to the consortium database to select its 50,000 names for mailing. These are the steps.

1. Draw a small sample of names ($N = 500$) from the consortium, mail catalogs to them, and see how much they purchase.
2. Perform a regression to determine the relationship between the available information for a customer and how much they spend.
3. Apply that regression model to the entire consortium database of names to predict spending levels.
4. Select the 50,000 names with the highest predicted spending levels.

The customer information variables are shown in this list. "Source" refers to the catalog company that supplied the name to the consortium.

source_a	Was the source company a? 1 = yes, 0 = no
source_b	Was the source company b? 1 = yes, 0 = no
source_r	Was the source company r? 1 = yes, 0 = no
Source_other	Was the source other? 1 = yes, 0 = no
Freq	How many orders in the last 2 years?
last_update_days_ago	How many days ago was customer record last updated (inquiry or order)?
Web order	Was order via web? 1 = yes, 0 = no
Address_is_res	Is address residential? 1 = yes, 0 = no

However, before we proceed, we need to stop and consider how categorical variables can be included in regression equations.

Binary and Categorical Variables in Regression

Until now, we have used only continuous variables in regression. As you have seen earlier, it is also possible to use binary—0 or 1—variables as independent or predictor variables, where a "1" or a "0" indicates a "yes" or "no" concerning the variable in question. The interpretation of the coefficient b in the resulting equation is simple. If the variable is present, for example, "address is residential," the outcome is increased by an amount equal to b as it is multiplied by 1. If the variable is not present, for example, "address is not residential," then the outcome is unaffected as b is multiplied by 0 and the term reduces to 0.

Example: Suppose we want to model how much a catalog customer will spend in the coming year and suppose the regression equation is

Spending $= 35 *$ gender $+ 0.0019 *$ income $+ 65$

Gender: 1 if female, 0 if male

Income: *Per capita* income in the state of residence

You can see from the regression equation that being female increases predicted spending by \$35. Let us say a customer comes from a state where the *per capita* income is \$40,000.

Spending if male: $35 * 0 + 0.0019 * 40{,}000 + 65 = \$141$

Spending if female: $35 * 1 + 0.0019 * 40{,}000 + 65 = \$176$

Where there are more than two categories for a variable, such as a "source" that could be one of three different companies, you need to create multiple binary "dummy" variables. You have seen this earlier, where source_a, source_b, and source_r are all listed as individual yes or no (0 or 1) variables. In other words, instead of a multicategory variable indicating "source could be a, b, or r," we have a series of binary variables indicating "is source = a?," "is source = b?," and so on. Note that a binary variable is also a categorical variable, one with only two categories.

Definition: **Dummy variable**

A dummy variable is a derived variable created by taking the categories in a categorical variable and making separate binary (yes/no) variables for each of those categories.

Multicollinearity

Multicollinearity is another issue to be aware of when bringing multiple variables into the picture.

Definition: **Multicollinearity**

Multicollinearity in multivariate modeling is the presence of one or more predictor variables that can be expressed as a linear combination of other predictor variable(s).

In this example, there are four possible sources, including "other."

However, suppose that we had only two possible sources—"source a" and "source b." And suppose that every name comes from one source or the other. So the variable "source_b" tells us nothing new—its information is exactly the same as the information contained in "source_a." If a name comes from "source a," it cannot come from "source b," and *vice versa*.

Similarly, if we have four possible sources—"a," "b," "r," and "other"—and every name has one and only one source, then once we know the values for "a," "b," and "r," the value for "other" is predetermined. If a source is "a," "b," or "r," the value for "other" must be 0, and if a source is not "a," "b," or "r," the value for "other" must be 1.

In both of the cases described earlier, we have multicollinearity: the information in one of the variables exactly duplicates the information contained in the other(s). This is a problem in regression because the mathematical operations needed to calculate the regression crash in the presence of multicollinearity. Even if variables are only approximately, rather than exactly, duplicative, the regression calculations can be unstable and unreliable.

Multicollinearity can happen with both continuous and categorical—including binary—variables. To avoid multicollinearity, two checks are needed.

1. When creating binary dummy variables from a single categorical variable, use a maximum of $k - 1$ dummies, where k is the number of categories.
2. When making decisions about which variables to include in a regression, consider whether they might be measuring the same thing and check the correlation between the two variables. Where correlation is high, you should use only one of them.

For calculation purposes, you will get the same results no matter which of the multi-collinear variables you omit, so you should retain the variables that are most informative. For example, in the problem described earlier, it would make sense to retain the source variables that contain the specific information and omit "other."

Tayko—Building the Model

Let us return to our goal, which is to build a model that predicts how much an individual will spend from a catalog. Our first step is to build a regression model using the data for which we have known spending outcomes. To avoid multicollinearity problems, we use only the three specific source categories, leaving out "other." Table 13.2 shows the first few rows of the data.

The regression output is given in Figure 13.14.

Reviewing the Output

The regression equation is

```
spending = 93.07 + 42.76(source_a) + 1.51(source_b) + 56.28(source_r) +
81.20(freq) - 0.66(last_update_days_ago) - 2.46(web_order)
```

Make sure you understand how you can derive this equation by looking at the software output.

We can see that the source catalogs "a" and "r" add to the spending estimate, by about $43 and $56, respectively. Each additional order (freq) in the customer's history adds about $81 to spending. Each additional day since the last contact (update) cuts the spending estimate by $0.66. *P*-values are all low for these variables. Catalog "b" as source, and whether an order comes from the web, seem to make little difference (coefficients are small and *p*-values quite high).

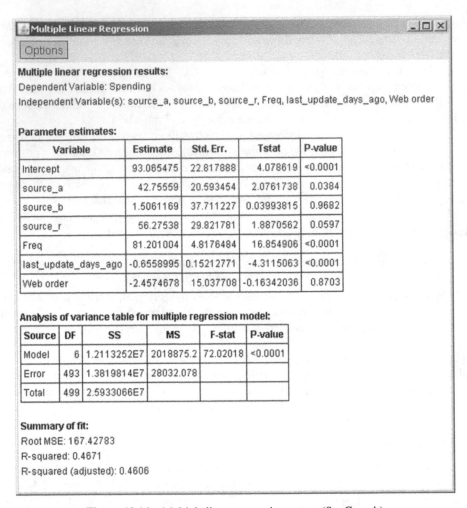

Figure 13.14   Multiple linear regression output (StatCrunch).

Predicting New Data

A brief review of the model coefficients and their significance is useful, but our real goal is to see how the model does in predicting spending when applied to new data.

The first few rows of the data to be predicted are shown in Table 13.3.

The goal is to fill in the unknown values with your predictions. How? Just apply the regression model that we have already developed:

```
spending = 93.07 + 42.76(source_a) + 1.51(source_b) + 56.28(source_r) +
81.20(freq) - 0.66(last_update_days_ago) - 2.46(web_order)
```

For the first row shown earlier, this works out as follows:

```
spending = 93.07 + 42.76(0) + 1.51(1) + 56.28(0) +
81.20(3) - 0.66(148) - 2.46(1) = $238.04
```

Doing so for the remainder of the data is left as a homework exercise.

TABLE 13.2   Tayko Data, Spending Known

A	B	C	D	E	F	G
source_a	source_b	source_r	Freq	last_update_days_ago	Weborder	Spending
0	1	0	2	183	1	128
0	0	0	2	194	0	127
1	0	0	1	161	0	174
0	0	0	1	73	0	192
0	0	1	2	147	1	386
0	0	0	2	73	0	174
0	0	0	1	123	1	189
0	0	0	1	165	0	90
0	0	0	2	147	1	352
0	0	1	9	43	1	639
0	0	0	2	89	0	41
1	0	0	$$$8	5	1	$$$1020

Question 13.2

We could also have used a holdout sample in this problem. Explain what its role would have been.

13.8   EXERCISES

1. Consider the following three hypothetical correlation matrices for airline data by city-pair, nonstop flights.

#1

	Fuel	Traffic	Price	Distance
Fuel	1			
Traffic	−0.6	1		
Price	0.65	−0.45	1	
Distance	0.75	0.25	0.55	1

TABLE 13.3   Tayko Data, Spending Unknown

A source_a	B source_b	C source_r	D Freq	E last_update_days_ago	F Weborder	G Spending
0	1	0	3	148	1	
0	0	0	3	145	0	
0	0	0	1	45	1	
0	0	0	6	70	0	
1	0	0	4	24	1	
0	0	0	6	74	0	
1	0	0	1	156	1	
0	0	0	2	27	1	
1	0	0	1	125	0	
0	0	0	1	105	1	
0	0	0	1	56	0	
0	0	0	1	145	0	
0	0	0	1	186	0	

#2

	Fuel	Traffic	Price	Distance
Fuel	1			
Traffic	−0.6	1		
Price	0.65	0.55	1	
Distance	0.90	0.25	0.75	1

#3

	Fuel	Traffic	Price	Distance
Fuel	1			
Traffic	0.25	1		
Price	0.65	−0.45	1	
Distance	0.90	0.25	0.75	1

Fuel = average fuel consumed per passenger
Traffic = number of passengers last year on the route
Price = average fare per passenger
Distance = miles between cities

Which of the hypothetical correlation matrices best fits the following background information?

Background: Aircraft use considerable fuel reaching cruising altitude, generally within 20–30 min of takeoff. Maintaining cruising altitude and speed require less fuel. Heavily trafficked routes tend to attract competition, which affects pricing. Airlines also use larger aircraft on heavily trafficked routes, and they are more fuel-efficient on a per person basis. Fares are higher for longer routes, in general, but among medium-haul and short-haul routes, there is less of a connection between distance and fares.

2. This problem uses an expanded version of the Boston Housing data. This version of the data contains five predictor variables for MEDV. The five variables are:

```
CRIM—crime rate per 1000 persons
NOX—nitric oxide concentration (parts per 10 million)
RM—average number of rooms per dwelling
PTRATIO—pupil-teacher ratio by town
LSTAT—% lower status of the population
```

   (a) Make a scatterplot of MEDV versus CRIM. What do you see? On the basis of your scatterplot, does CRIM appear helpful in predicting MEDV?

   (b) Run a regression of MEDV as a function of CRIM. Report a *p*-value and interpret it. From this, does CRIM appear helpful in predicting MEDV?

   (c) Make a scatterplot of MEDV versus LSTAT. What do you see? On the basis of your scatterplot, does LSTAT appear helpful in predicting MEDV?

   (d) Run a regression of MEDV as a function of LSTAT. Report a *p*-value and interpret it. From this, does LSTAT appear helpful in predicting MEDV?

   (e) So far, we have been analyzing one predictor variable at a time so we do not have a good idea how they work together to affect MEDV. Run a multiple regression of MEDV versus all five of the predictor variables. Report the regression equation and the predictor variable *p*-values from the computer output. From this, does the regression appear helpful in predicting MEDV?

3. Use the dataset "Tayko-known.xls" for the following problems.

   (a) Perform a multiple linear regression. Specify appropriate independent and response variables. Report the resulting equation. (Hint: this duplicates an illustration in the chapter.)

   (b) Calculate, or locate in your regression output, the following statistics, and use them in a sentence or two describing the Tayko-known data and your regression. The goal is to convey to your reader an understanding of average spending, how variable it is, and how much a typical prediction might be in error.
   – standard deviation of the spending
   – mean spending
   – RMSE

   (c) Use the regression equation to predict spending levels for the customer records in "Tayko-Unknown.xls."

   (d) Sort the results in #3 by predicted spending, and report the top 10 customers for predicted spending.

(e) Your goal is to generate at least $250 in spending per catalog mailed. How many customers should you mail to, from Tayko-unknown.xls? (Hint: would you mail a catalog to the customer represented by the top row in part (d)? The second row? The last row?

Note: (b) and (c) do not require statistical software and can be done in both StatCrunch and Excel. StatCrunch will calculate the RMSE (Root MSE) automatically with a regression equation.

4. The following regression equations relate to concentrations of chlorophyll (a measure of lake quality—higher levels indicate algae and possible eutrophication), phosphate, and nitrogen. (Based partially on a problem in Manly, B., *Statistics for Environmental Science and Management*, 2nd ed., CRC Press, p.71)
   The specification of the model by the researcher:

```
(1) CH = b0 + b1(PH) + b2(NT)
The regression output:
(2) CH = -9.386 + 0.333PH + 1.200NT
Output from an additional regression:
(3) CH = -16.244 + 0.313PH + 0.960NT + 0.412(NT)(PH)
```

Questions (answer each with single short sentence, or even shorter):

(a) Considering the three variables and how they are arranged in Equation (1), what does the researcher believe is the "cause and effect" relationship among the three variables (i.e., what causes what)?

(b) What is different about (2) and (3)?

(c) What is the effect of increasing NT by 1 unit in Equation 2 and keeping PH fixed?

(d) What is the effect of increasing NT by 1 unit in Equation 3 and keeping PH fixed?

Answers to Try It Yourself

13.1 Without any antimony, the alloy strength is predicted to be 20.2 units. Strength increases by 0.087 units for each percentage point increase in antimony.

13.2 See file rmse_example.xls

13.3 When RM = 5, a 1-unit increase in CRIM yields a change in MEDV of $1.247 - (0.2916 * 5) = -0.211$, or a loss in the value of $211. When RM = 7, a 1 unit increase in CRIM yields a change in MEDV of −0.794, or a loss in the value of $794. For the vast majority of houses, the net effect of CRIM is negative, not positive. The two regression terms—the individual CRIM coefficient and the interaction coefficient—work together so that the negative impact of CRIM has more effect with larger houses.

Answers to Questions

13.1 The y-intercept is the predicted value of the outcome variable when all the predictor variables are 0. Houses with 0 rooms do not exist so we are not concerned with the y-intercept.

13.2 Instead of running the regression on all the available data where the spending was known, we would have set aside a portion (generally 30–50%) as a holdout sample and not used it in fitting the model. Once the model was developed with the training data (the data not taken for the holdout sample), we would apply the resulting regression equation to the holdout data, and calculated predicted values. We would then have compared predicted to actual in the holdout data and calculated RMSE to provide an unbiased estimate of how well the regression would perform with new data.

INDEX

absolute residual error, 212
alpha, 53, 153

Bayesian calculations, 96
Benford's Law, 187
Bernoulli trials, 146
best fit, 210
bias, 12, 106
Big Data, 27
Binary Data, 15
Binary Variables in Regression, 274
binomial distribution, 134, 147
blocking, 247
bootstrap, 114, 126
    parametric 141
bootstrap confidence interval for a regression
        coefficient, 265
Bootstrapping a Regression, 264
box, 111
Box Sampler, 4
boxplot, 37

categorical, 29, 59
causation, 205
Central Limit Theorem, 129
chi-square, 186
cluster sampling, 118
coincidence, 205
complement, 74
conditional probability, 92
confidence interval, 125

contingency table, 182
control group, 8
convenience sampling, 118
correlation, 193
correlation coefficient, 199
correlation matrix, 258
critical value, 154
cumulative frequencies, 34

data analytics, 28
database normalization, 25
degrees of freedom, 19
deviations, 17, 52
difference in proportions,
double-blind, 12
Dummy Variable, 274

error, 211

factorial design, 246
flat file, 25
frequency distribution, 33
frequency table, *see* frequency distribution, 33

Gallup, 106
GDP, 62
Google Trends, 66

hat, *see* box 111
Hawthorne effect, 12
hinges, 17

Introductory Statistics and Analytics: A Resampling Perspective, First Edition. Peter C. Bruce.
© 2015 John Wiley & Sons, Inc. Published 2015 by John Wiley & Sons, Inc.

histogram, 35
hypothesis test, 45

independent events, 98
interaction, 242
interquartile range, 17
IQR, *see* interquartile range, 17

law of averages, 9
law of large numbers, 9
least squares regression, 212
linear regression, 212
linear relationship, 213

margin of error, 109, 125
marginal columns and rows, 89
mean, 13
measurement variable, 29
median, 13
Mendel's peas, 72
mode, 15
multicollinearity, 275
multistage sampling, 118
munging, 24

nonresponse bias, 119
normal distribution, 52, 80
normalization, 25
nuisance variable, 248
null hypothesis, 155
null model, *see* null hypothesis, 152, 155

observation, 113
observational study, 7
outlier, 32

paired data, 13
parameter, 108
parametric bootstrap, 141
percentile, 17
permutation test, 46
pie charts, 61
PivotTable, 92
placebo effect, 12
point estimate, 125
population, 108
power, 165
practical significance, 55
prior probability, 96
probability, 73
probability distribution, 74, 78
pseudo-random number generator, 7
p-value, 153

qualitative variable, *see* categorical variable, 29

R_2, 261
random assignment, 8

random number, 7
random number generator, 7, 141
random sample, 108
    simple 107
random variable, 77
range, 16
regression, 209
regression line, 211
Relational Database, 25
relative frequencies, 34
replication, 225
resample, 113
resample with replacement, *see* sample with replacement, 114
residual, 211
RMSE, 262
root mean squared error, 262

sample, 108, 113
sample standard deviation, 19
sample variance, 19
sample with replacement, 111
sampling error, 118
sampling frame, 108, 118
sampling with replacement, 114
sampling without replacement, 114
self selection, 119
shuffling, 114
significance level, 153, 229
simple random sample (SRS), *see* random sample (simple)
Simpson's Paradox, 89
simulated population, 111, 144
simulation, 114
single simulation trial, 114
single-blind, 12
skew, 39
slope coefficients, 217
SQL, 26
squared residual error, 212
SRS, *see* random sample (simple), 107
standard deviation, 18
standard error, 132
statistic, 109
statistical significance, 55, 153
STDEV, 19
stem-and-leaf plot, 36
stratification, 247
stratified sampling, 117
Structured Query Language, 26
Student's t, 131
survey, 104
systematic sampling, 118

t-distribution, 131
tail of the distribution, 39
test statistic, 22
treatment group, 8

trend line, 210
triple-blind, 12
Tukey, 99
Type I error, 54
Type II error, 54

urn, *see* box, 111

VAR, 19

variability, 16
variable, 77
variance, 18
Venn Diagrams, 74

whiskers, 37

z-interval, 130